A Maple Supplement for Linear Algebra

John Maloney
University of Nebraska at Omaha

Upper Saddle River, NJ 07458

Editor-in-Chief: Sally Yagan
Executive Editor: George Lobell
Supplement Editor: Jennifer Urban
Executive Managing Editor: Kathleen Schiaparelli
Assistant Managing Editor: Becca Richter
Production Editor: Allyson Kloss
Supplement Cover Manager: Paul Gourhan
Supplement Cover Designer: Joanne Alexandris
Assistant Manufacturing Manager: Michael Bell
Manufacturing Buyer: Ilene Kahn

Pearson Prentice Hall
Pearson Education, Inc.
Upper Saddle River, NJ 07458

Printed in the United States of America

10 9 8 7 6 5 4 3 2 1

ISBN 0-13-145337-8

Pearson Education Ltd., *London*
Pearson Education Australia Pty. Ltd., *Sydney*
Pearson Education Singapore, Pte. Ltd.
Pearson Education North Asia Ltd., *Hong Kong*
Pearson Education Canada, Inc., *Toronto*
Pearson Educación de Mexico, S.A. de C.V.
Pearson Education—Japan, *Tokyo*
Pearson Education Malaysia, Pte. Ltd.

Chapter 1 Getting Started With Maple

1. Basic operations
2. Some useful maple functions
3. Calculus based operations
4. Creating and using functions
5. Do loops and if--then--else
6. Plotting

Chapter 2.0 Basic Operations using the linalg package

1. Basic Matrix Operations
2. Important Vector Concepts
3. Vector Projection
4. Graphs and Incidence Matrices
5. Markov Chains
6. Complex Values

Projects for chapter 2

Project 1
Project 2
Project 3
Project 4

Chapter 2.1 Basic Operations using the LinearAlgebra package

1. Basic Matrix Operations
2. Important Vector Concepts
3. Vector Projection
4. Graphs and Incidence Matrices
5. Markov Chains
6. Complex Values

Chapter 3.0 Solving Systems of Equations using the linalg package

1. Equation Reduction and the Augmented Matrix
2. Reduction to Echelon Form
3. Singular systems
4. The inverse Matrix and the LU decomposition
5. Linear Independence, Rank, and Determinant

Projects for chapter 3

Project 1
Project 2
Project 3
Project 4

4. **Project 4**

Chapter 5.1 Linear Transformations using the LinearAlgebra package

1. **Linear independence and span**
2. **Subspaces and basis**
3. **Linear transformations**
4. **Change of basis.**

PREFACE

Maple is a computer algebra software package that is capable of doing an amazing amount of mathematical computation. With the exception of the first chapter this text explores only its capability in solving linear algebra problems. It is fair to say that these relatively new computer algebra software packages have changed the way that mathematics courses can be taught. Large problems and problems that are very computationally intensive can now be incorporated into courses. Such problems can now be incorporated into courses by having maple (and similar programs) handle the extensive computation.

The first chapter provides a brief introduction showing how Maple can be used to solve ordinary calculus problems. There are two packages in Maple for doing linear algebra, they are the **linalg** package and the **LinearAlgebra** package. The syntax is considerably different for each package. For this reason each chapter (after the first one) is repeated. First the material is discussed using the **linalg** package. Next exactly the same material and examples are discussed using the syntax for the **LinearAlgebra** package. Thus chapter 2 would use the syntax for the **linalg** package while chapter 2.1 would use the syntax for the **LinearAlgebra** package. Other than the Maple syntax the contents of chapters 2 and 2.1 are identical. Similarly for the other chapters.

The author wishes to dedicate this book to the memory of his parents John and Helen Maloney.

John Maloney
University of Nebraska at Omaha

Chapter 1
Getting Started With Maple

1. Basic Operations

The purpose of this book is to show how to use the software package Maple to solve many of the problems that arise in the study of Linear Algebra. Software packages such as maple have taken a great deal of the drudgery out of doing mathematics. Their ability to perform mathematical and symbolic calculations is truly amazing.

We begin with some simple examples illustrating the basics of arithmetic.

```
> 3+5-4;
```

$$4$$

```
> 3*5+5*6^3;
```

$$1095$$

```
> 2^(-5);
```

$$\frac{1}{32}$$

Maple has all of the standard mathematical functions, such as sin, cos, e, ln, etc., built into it.

```
> z:=15*sin(Pi/4)+13*arctan(-43);
```

$$z := \frac{15\sqrt{2}}{2} - 13\arctan(43)$$

Note that maple has expressed the result in terms of the arctan and radicals rather than as a number. The reason being that this form of the solution is more accurate than a 10 digit approximation. Oftentimes we want the number even though it may be less accurate. The maple function **evalf** will force the result down to a single number.

```
> w1:=evalf(z);
```

$$w1 := -9.51147943$$

```
> q:=3*exp(-0.75)+11*ln(3^3-5);   evalf(%);
```

$$q := 1.417099658 + 11\ln(22)$$

$$35.41856664$$

We should note that in maple π is **Pi** and e^x is **exp(x)**. All of the standard mathematical functions such as the trig, hyperbolic, and all of the others as well, are available in maple. In maple we can have as many digits, in the answer, as we could reasonably desire. To change the number of digits use:

```
> Digits:=25;
```

$$Digits := 25$$

```
>q:=3*exp(-0.75)+11*ln(3^3-5);    evalf(%);
```

$$q := 1.41709965822304412141414 0 + 11\ \ln(22)$$

$$35.41856664516451850968508$$

```
>Digits:=10; # 10 is the default number of digits
```

2. Some useful maple functions

In this section we shall look at several maple functions that can be used to simplify and/or change a given expression into another form. The first two functions that we want to look at are **expand** and **factor** which are, as their names imply, opposites.

```
>p:=(2*x+5)*(x^2-3)*(3*x^3+7);
```

$$p := (2\,x+5)\,(x^2-3)\,(3\,x^3+7)$$

```
>p1:=expand(p);
```

$$p1 := 6\,x^6 - 31\,x^3 - 18\,x^4 - 42\,x + 15\,x^5 + 35\,x^2 - 105$$

Note that the polynomial is not written in terms of descending powers of x. To obtain this result we shall use the **sort** function.

```
>p2:=sort(p1);
```

$$p2 := 6\,x^6 + 15\,x^5 - 18\,x^4 - 31\,x^3 + 35\,x^2 - 42\,x - 105$$

To put this expression back into factored form we can use:

```
>p3:=factor(p2);
```

$$p3 := (2\,x+5)\,(x^2-3)\,(3\,x^3+7)$$

Oftentimes we would like to combine several rational expressions together by putting them over a common denominator. This can be easily accomplished with the **normal** command.

```
>p:=(2*x^3+3)/(5*x-7)+17/(3*x^2+5);
```

$$p := \frac{2\,x^3+3}{5\,x-7} + \frac{17}{3\,x^2+5}$$

```
>p1:=normal(p);
```

$$p1 := \frac{6\,x^5 + 10\,x^3 + 9\,x^2 - 104 + 85\,x}{(5\,x-7)\,(3\,x^2+5)}$$

We might like to try and rearrange this by either factoring the numerator or by expanding the denominator. To pick out the numerator and denominator we use the **numer** and **denom** commands provided by maple.

```
> p2:=numer(p1);   p3:=denom(p1);
```

$$p2 := 6\,x^5 + 10\,x^3 + 9\,x^2 - 104 + 85\,x$$

$$p3 := (5\,x - 7)\,(3\,x^2 + 5)$$

```
> p21:=factor(p2);
```

$$p21 := 6\,x^5 + 10\,x^3 + 9\,x^2 - 104 + 85\,x$$

The above result shows that maple cannot factor p2.To get more explicit information on the factor (or any other maple) command look up the help page by using ?command. In this case we would use **?factor** and after a carriage return maple will bring up the help page. The help page begins with an explanation of the command and its parameters, if any, and is followed by several examples.

Two more useful commands are **simplify** and **combine** which can be used to simplify algebraic and trigonometric expressions.

```
> p:=5*sin(3*x)^2+5*cos(3*x)^2+3*tan(5*x);
```

$$p := 5\,\sin(3\,x)^2 + 5\,\cos(3\,x)^2 + 3\,\tan(5\,x)$$

```
> p1:=simplify(p,trig);
```

$$p1 := \frac{80\,\cos(x)^5 - 100\,\cos(x)^3 + 48\,\sin(x)\,\cos(x)^4 - 36\,\sin(x)\,\cos(x)^2 + 3\,\sin(x) + 25\,\cos(}{\cos(5\,x)}$$

This is not what we had anticipated so lets use the other command

```
> p1:=combine(p);
```

$$p1 := 5 + 3\,\tan(5\,x)$$

This is what we had expected to get. Below, note the use of **exp()** for the exponential function.

```
> p:=exp(a+ln(x^2+2))+4^(3/2)  +8*x;
```

$$p := \mathrm{e}^{(a + \ln(x^2 + 2))} + 4\,\sqrt{4} + 8\,x$$

```
> p1:=simplify(p);
```

$$p1 := \mathrm{e}^a\,x^2 + 2\,\mathrm{e}^a + 8 + 8\,x$$

```
> p2:=combine(p);
```

$$p2 := (x^2 + 2)\,\mathrm{e}^a + 4\,\sqrt{4} + 8\,x$$

Now lets look at an application of the **collect** command. As its name implies it collects things, namely coefficients of x or y or …

```
> p:=a*x^2+b*y^2+c*x^2+d*y^2+k*x+m*y+3*x+5*y+6;
```

$$p := a\,x^2 + b\,y^2 + c\,x^2 + d\,y^2 + k\,x + m\,y + 3\,x + 5\,y + 6$$

```
> p1:=collect(p,[x,y]);
```

$$p1 := (c+a)\,x^2 + (3+k)\,x + (b+d)\,y^2 + (m+5)\,y + 6$$

Exercises

1). Expand the following functions using maple's **expand** function. In the case of polynomials use the **sort** function to sort them into proper order.

a). $f(x) = (2x+4)(7x-2)(17x+19)$.

b). $f(x) = (5x-3)^3(12x+15)^2$.

c). $f(x) = 3\cos(2x) - 5\sin(3x)$.

d). $f(x) = \cos(5x)$.

2). Use maple's factor function to factor the following polynomials.

a). $f(x) = x^2 - 5x - 6$.

b). $f(x) = x^3 + 4x^3 + 4x$.

3). Use the maple functions studied in section 2 to simplify as much as possible the following expressions.

a). $f(x) = \dfrac{2x+5}{x^2+2} + \dfrac{7x^2-9}{3x-4} + \dfrac{22x+17}{x^2-2}$.

b). $f(x) = 3\cos^2(5x) + 3\sin^2(x) + 3\tan^2(2x)$.

c). $f(x) = 2 + 2\tan^2(x) + x^2 + 6x - 5 + (2x-5)(x+8)$.

3. Calculus based operations

The most common operations in calculus are those of integrating and differentiating functions. For these operations we use the maple commands **int** and **diff.** Some examples are provided below:

```
> ex1:=(5*x^2+3*x+17)*cos(19*x)^2;
```

$$ex1 := (5\,x^2 + 3\,x + 17)\cos(19\,x)^2$$

```
> ex11:=int(ex1,x);
```

$$ex11 := \frac{5}{19}x^2\left(\frac{1}{2}\cos(19\,x)\sin(19\,x) + \frac{19\,x}{2}\right) + \frac{5}{722}\cos(19\,x)^2\,x$$
$$+ \frac{12269}{27436}\cos(19\,x)\sin(19\,x) + \frac{12269\,x}{1444} - \frac{5\,x^3}{3} + \frac{3}{19}x\left(\frac{1}{2}\cos(19\,x)\sin(19\,x) + \frac{19\,x}{2}\right)$$
$$- \frac{3}{1444}\sin(19\,x)^2 - \frac{3\,x^2}{4}$$

```
> ex2:=(5*x^2+3*x+17)*cos(19*x)^2;
```

$$ex2 := (5\,x^2 + 3\,x + 17)\cos(19\,x)^2$$

```
> ex21:=int(ex2,x=a..b);
```

$$ex21 := \frac{5}{38}b^2\cos(19\,b)\sin(19\,b) + \frac{5\,b^3}{6} + \frac{5}{722}\cos(19\,b)^2\,b + \frac{12269}{27436}\cos(19\,b)\sin(19\,b)$$
$$+ \frac{12269\,b}{1444} + \frac{3}{38}b\cos(19\,b)\sin(19\,b) + \frac{3\,b^2}{4} - \frac{3}{1444}\sin(19\,b)^2$$
$$- \frac{5}{38}a^2\cos(19\,a)\sin(19\,a) - \frac{5\,a^3}{6} - \frac{5}{722}\cos(19\,a)^2\,a - \frac{12269}{27436}\cos(19\,a)\sin(19\,a)$$
$$- \frac{12269\,a}{1444} - \frac{3}{38}a\cos(19\,a)\sin(19\,a) - \frac{3\,a^2}{4} + \frac{3}{1444}\sin(19\,a)^2$$

```
> ex22:=int(ex2,x=0..Pi); evalf(%);
```

$$ex22 := \frac{5}{6}\pi^3 + \frac{3}{4}\pi^2 + \frac{12279}{1444}\pi$$

$$59.95518286$$

```
> ex2:=Int(ex2,x=a..b);
```

$$ex2 := \int_a^b (5\,x^2 + 3\,x + 17)\cos(19\,x)^2\,dx$$

The above examples indicate how we can use the **int** command to do either definite or indefinite integration. In the last case we have that **Int** simply reproduces the standard symbolic mathematical notation for the integral to be found. It does not do any computation. Many integrals do not have closed form solutions and a numerical technique must be used to find the value. For these we use the format **evalf(Int(f(x),x=a..b))**. For example:

```
> ex1:=evalf(Int(exp(x^2+3),x=0..7));
```

$$ex1 := 0.2765258914\;10^{22}$$

```
> ex2:=evalf(Int(cos(x^2+5),x=0..Pi/2));
```

$$ex2 := 1.034969602$$

For differentiation we have

```
> f:=(x^2+3*x-17)*tan(5*x)+sqrt(x^2+7);
```

$$f := (x^2 + 3\,x - 17)\tan(5\,x) + \sqrt{x^2+7}$$

```
> f1:=diff(f,x);    # first derivative
```

$$f1 := (2\,x + 3)\tan(5\,x) + (x^2 + 3\,x - 17)(5 + 5\tan(5\,x)^2) + \frac{x}{\sqrt{x^2+7}}$$

```
> f2:=diff(f,x$2); # second derivative
```

$$f2 := 2\tan(5x) + 2(2x+3)(5+5\tan(5x)^2) + 10(x^2+3x-17)\tan(5x)(5+5\tan(5x)^2) - \frac{x^2}{(x^2+7)^{(3/2)}} + \frac{1}{\sqrt{x^2+7}}$$

```
> f3:=diff(f,x$3);  # third derivative
```

$$f3 := 30 + 30\tan(5x)^2 + 30(2x+3)\tan(5x)(5+5\tan(5x)^2) + 10(x^2+3x-17)(5+5\tan(5x)^2)^2 + 100(x^2+3x-17)\tan(5x)^2(5+5\tan(5x)^2) + \frac{3x^3}{(x^2+7)^{(5/2)}} - \frac{3x}{(x^2+7)^{(3/2)}}$$

Exercise

1). In each case use **Int** to reproduce the mathematical formula and then compute the given integral. If possible compute the exact value. If this cannot be found then compute a numerical approximation to the integral.

a). $\int_1^{16}(x^3+15x)dx.$

b). $\int(x^2+5)dx.$

c). $\int_a^b \frac{x^2-5x+9}{(2x-3)(4x+5)(x^2+2)}dx.$

d). $\int_0^1 \frac{x^2-5x+9}{(2x-3)(4x+5)(x^2+2)}dx.$

e). $\int(5x^2+17x-19)\cos(11x)dx.$

2). Find the first, second, and third derivatives of the integrand functions in problem 1.

4. Creating and using functions

In many cases we would like to create and use our own functions. This will allow us to use standard mathematical notation. In this section we will show how to do this. We begin by showing how to evaluate expressions using the **subs** command.

```
> exp1:=3*x^2+5*x-17 +cos(x);
```

$$\text{exp1} := 3x^2 + 5x - 17 + \cos(x)$$

```
> exp1_val:=subs(x=Pi/4,exp1);    evalf(%);
```

$$exp1\_val := \frac{3\,\pi^2}{16} + \frac{5\,\pi}{4} - 17 + \cos\left(\frac{\pi}{4}\right)$$

$$-10.51535158$$

The advantage of using the **subs** command is that it leaves the expression exp1 unchanged. Now let us convert exp1 to a function so that we can evaluate it in the usual way. There are two ways to do this. One is by using the **unapply** function and the other is by creating the function directly.

```
> f1:=unapply(exp1,x);
```

$$f1 := x \to 3\,x^2 + 5\,x - 17 + \cos(x)$$

```
> f2:=x->3*x^2+5*x-17 +cos(x);
```

$$f2 := x \to 3\,x^2 + 5\,x - 17 + \cos(x)$$

```
> f1(Pi/4);   f2(Pi/4);
```

$$\frac{3}{16}\pi^2 + \frac{5}{4}\pi - 17 + \frac{1}{2}\sqrt{2}$$

$$\frac{3}{16}\pi^2 + \frac{5}{4}\pi - 17 + \frac{1}{2}\sqrt{2}$$

Maple expresses the answer in this form because it is the most accurate form. Replacing Pi by a ten digit approximation would be less accurate. Never the less, there are times when we want a single floating point answer. To get it we use the **evalf** command which forces the computation down to a single floating point answer.

```
> evalf(f1(Pi/4));        evalf(f2(Pi/4));
```

$$-10.51535158$$

$$-10.51535158$$

Note that both methods give the same result. Note also that the functions f1 and f2 are evaluated by using the standard mathematical notation rather than the maple command **subs**.

For differentiating functions such as f1 or f2 we should use **D** (see the help page **?D**) as shown below. The reason for this is that the result is another function. This is very useful, since the original is a function, the most useful form of the derivative would be a function.

```
> f1_1:=D(f1);
```

$$f1\_1 := x \to 6\,x + 5 - \sin(x)$$

```
> f1_2:=D[1$2](f1);   # second derivative
```

$$f1\_2 := x \to 6 - \cos(x)$$

```
> f1_3:=D[1$3](f1);   # third derivative
```

$$f1\_3 := \sin$$

```
> f1_1(2);    f1_2(2);   f1_3(2);
```

$$17 - \sin(2)$$

$$6 - \cos(2)$$

$$\sin(2)$$

Functions can be combined in the usual mathematical way.

```
> w1:=evalf((f1+f2)(Pi));
```

$$w1 := 54.63355296$$

```
> w1:=evalf((f1*f2)(Pi));
```

$$w1 := 746.2062773$$

Exercises

1). Use maple to create each of the following functions and then evaluate them (to a single number), using **evalf** where necessary, at the points x = 1, 5, and -17.

a). $f(x) = (4x - 5)(x^3 + 17x)$.

b). $f(x) = \dfrac{7x + 19}{x^3 + 20}$.

2). Same as in 1 but evaluate the functions and their first two derivative functions

a). $f(x) = 2\cos(3x) + 5\sin(2x) - \tan(x)$.

b). $f(x) = \dfrac{\cos^2(8x) + 5}{\sin^2(3x) + 7}$.

5. **Do loops and if--then--else**

Do loops are a powerful feature of maple. If you are already familiar with do loops from other programming languages you will find that they work pretty much the same way in maple. You should look up the help page **?do**. There are several versions and we shall look at examples of them all.

Example 1.1

Compute the sum of the first 15 positive integers.

```
> s:=0:    # We will acumulate the sum in s.
> for j from 1 to 15 do
>   s:=s+j;
> od:
> print(s);
```

In this loop the initial value of the loop counter j is one and the final value is 15. Each time through the loop j is indexed, that is increased, by one since no other index has been specified. You can specify other indices if you wish. We can think of the indexing as taking place at the bottom of the loop. After j is indexed at the bottom of the loop its value is compared to the limit, here 15, and if it does not exceed the limit then the loop is executed again using the new value of j. If the loop counter j exceeds the limit then we drop out of the loop and go to the print statement.

A few things to note are that the use of the colon in **s:** and **od:** eliminates unwanted output. The statements are executed but the output of the calculation is suppressed. Note that the value s = 0 was not printed out. In connection with do loops it is always a good policy to use nothing but **od:** to suppress unwanted output. If we wish to output any of the internal loop computations we should use an explicit **print** statement.

Example 1.2

Subtract the quantities 20, 18, 16, 14, and 12 from 480 and print out the result.

```
> s:=480:
> for j from 20 by -2 to 12 do
>   s:=s-j;
> od:
> print(j,s);
```

$$10, 400$$

In this loop j is decreased by 2 at the end of the loop since the index has been specifically given. Now lets redo a previous example and include a print statement within the loop to print out the values of j and s. Clearly we do not want to print out all of the values. We shall print out selected values using the **mod** function. The expression **w: = a mod 5;** assigns to w the remainder on dividing a by 5 which means that w can have any one of the values 0, 1, 2, 3, and 4.

Thus we can print when j mod 5 has the value 0, which means we will print when j = 5, 10, 15, ... Here we need to make use of the **if--then--else** statement which has the form

```
if (boolean conditions)
     then statement1;
          .
          .
          statementm;
     else statement1;
          .
          .
          statementk;
fi;
```

```
> s:=0:
> for j from 1 to 1000 while(s<185) do
>   s:=s+j;
>   if (j mod 5 = 0) or (s>=185)
>          then print(j,s);
>   fi;
> od:
```

$$5, 15$$

$$10, 55$$

$$15, 120$$

$$19, 190$$

The index, by default, is one. We now have the additional condition that $s < 185$. At the bottom of the loop j is indexed by one and if either $j > 15$ or $s \geq 185$ then the loop is terminated. Note that for $j = 5$, 10, and 15 the printout is due to the **mod** option while for $j = 19$ the printout is do to the condition s **>= 185**. The value that we are seeking is s-j that is $190 - 19 = 171$. In our later work we will see many examples of **do loops** and the **if--then--else** statement.

Exercises

1). Redo the above loop so that s contains the correct value when the loop terminates.

2). Find the sum of the first 15 even integers and print out the final result.

3). Find the sum of the first 40 odd integers and print out the final result.

4). Compute the sum $s = 1 + \frac{1}{2} + \frac{1}{3} + \cdots + \frac{1}{n} + \cdots$ up to the point where s exceeds 4.3. Print out the value of the loop counter and the current sum s every 10 times through the loop as well as the last time through the loop.

5). Same as problem 3 except you are to print out the values of the loop counter and the current sum at loop counter values of 5, 10, 15, etc.

6). Using the **if--then--else** to define a function f(x) with the property that

$$f(x) = \begin{cases} x^2 - 19 & \text{if } x \leq 6 \\ 7x - 15 & \text{if } x > 6 \end{cases}.$$

6. Plotting

We will need only the simplest plotting facilities (which are large) of maple for linear algebra. Let us begin first with a simple plot of a function and then of two functions.

```
> plot(sin(3*x),x=0..2*Pi);
```

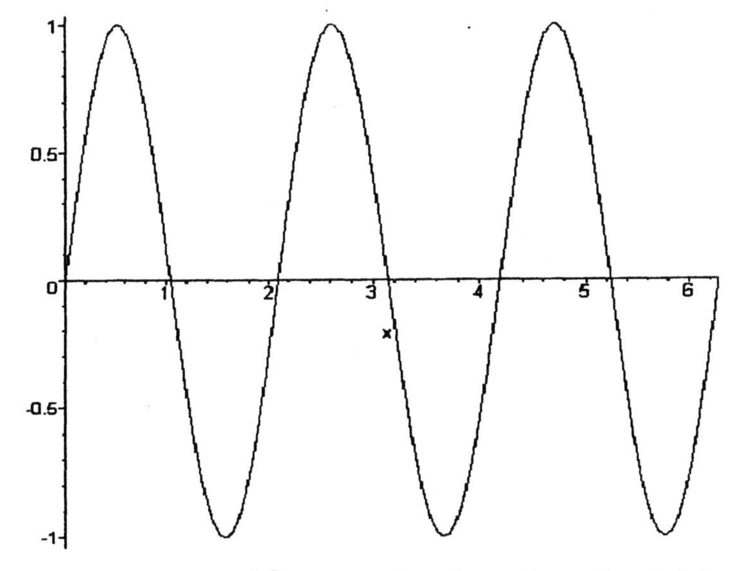


```
> plot({sin(3*x),cos(3*x)},x=0..2*Pi,color=[red,blue]);
```

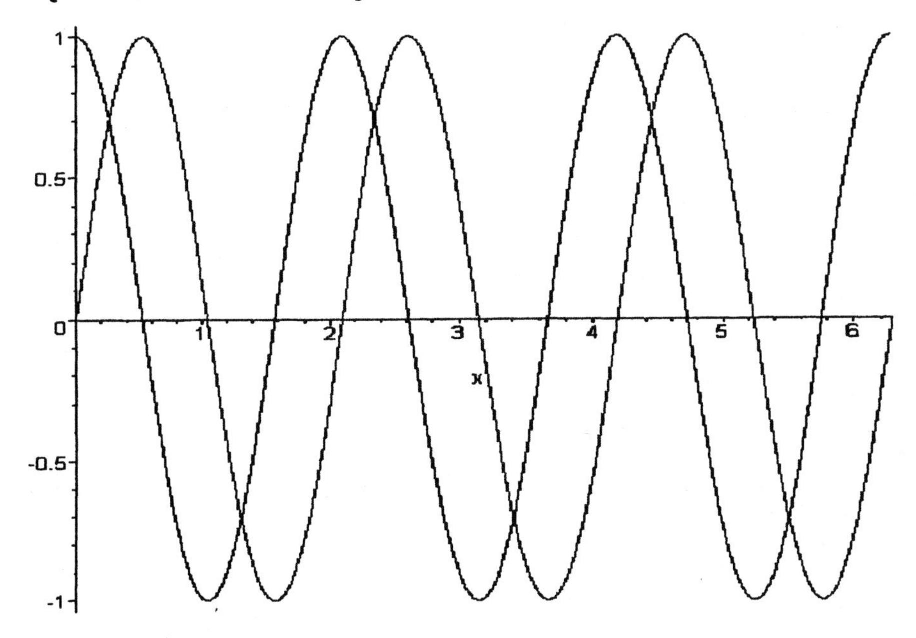

Note that to plot two fucntions we enclose them in braces {}. {} are used in maple to denote a set. Thus we have made a set of the two functions and asked maple to plot the set of functions. We have also specified the colors by means of the color option. At this point we should note that it is important to maple whether we use (), [], or {}as they are used for different purposes. We can put in a y range if we wish to restrict the y coordinates as follows:

```
> plot(x^2,x=-3..3,y=0..4);
```

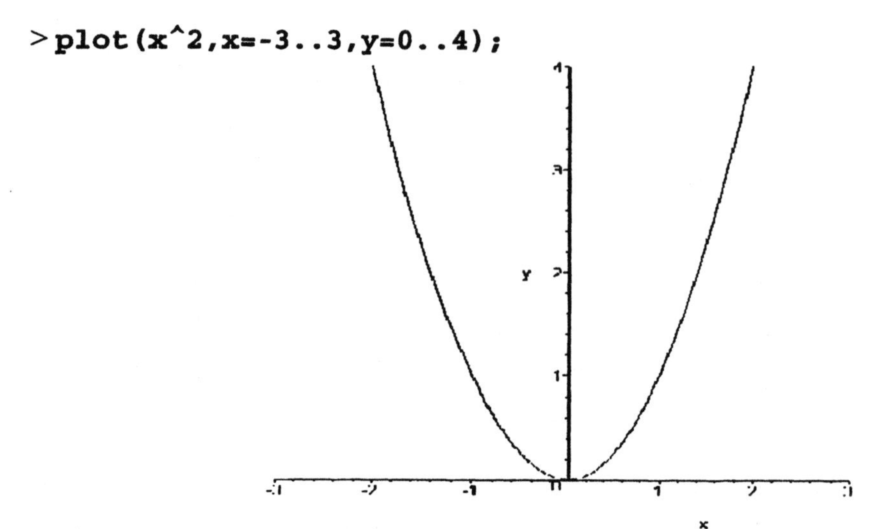

Note that the y coordinate of the plot does not exceed 4.

Next we shall look at plotting figures such as triangles, squares, circles etc. We shall specify the vertices that are to be connected by straight lines. We shall make a list of the coordinates of the 3 vertices of the triangle and another for the square. In maple a list is any collection of objects, separated by commas, and contained between the braces []. Note that the first listed vertex also appears at the end of the list. The reason for this is that maple will connect each vertex, in succession, with the next. To complete the figure the last vertex must be connected to the first. The basic plotting functions are read into maple when the program is loaded into memory but the advanced features that we will be using are not. They are contained in a package called **plots.** We shall now read in the plots package.

```
> with(plots):
Warning, the name changecoords has been redefined
```

```
>T1:=[[1,1],[4,1],[2,5],[1,1]]; # Note that the first
                                  # pointis repeated at the end.
```

$$T1 := [[1, 1], [4, 1], [2, 5], [1, 1]]$$

```
> S1:=[[0.5,0.5],[5,0.5],[5,7],[0.5,7],[0.5,0.5]];
```

$$S1 := [[0.5, 0.5], [5, 0.5], [5, 7], [0.5, 7], [0.5, 0.5]]$$

```
> listplot(T1);
```

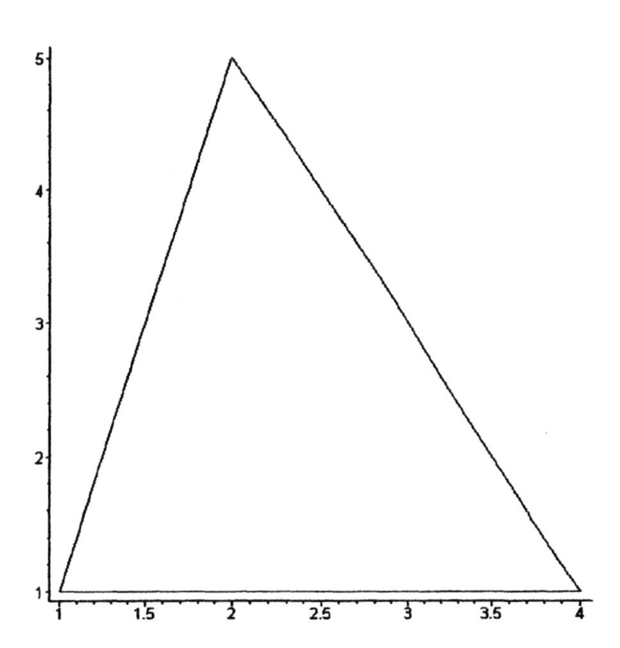


```
> listplot(S1);
```

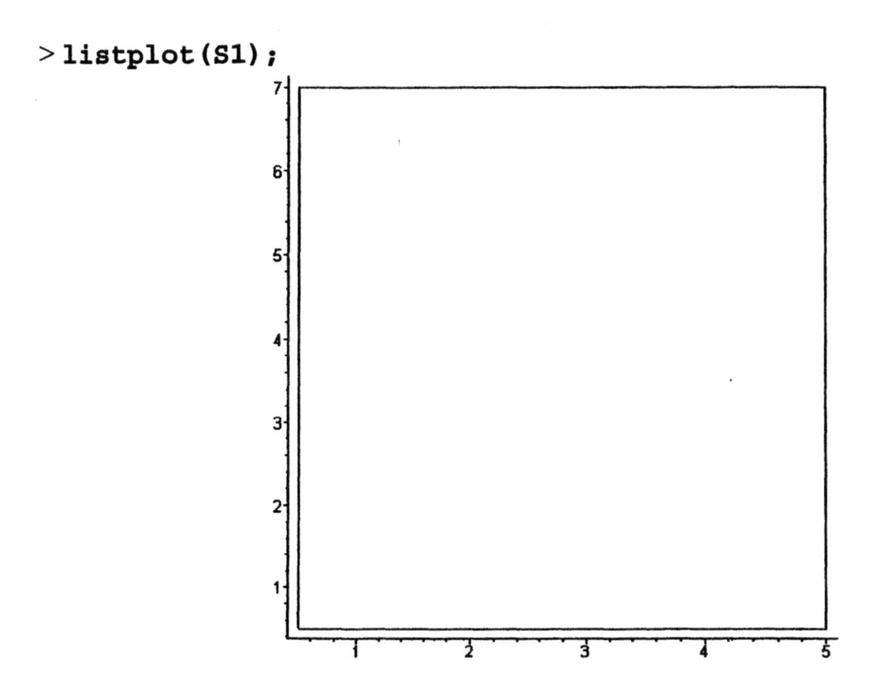

We have plotted each function separately. How can we combine the two plots together? The easiest way to do this is to use the **display** functions.

```
> P1:=listplot(T1):
> P2:=listplot(S1):
> display(P1,P2);
```

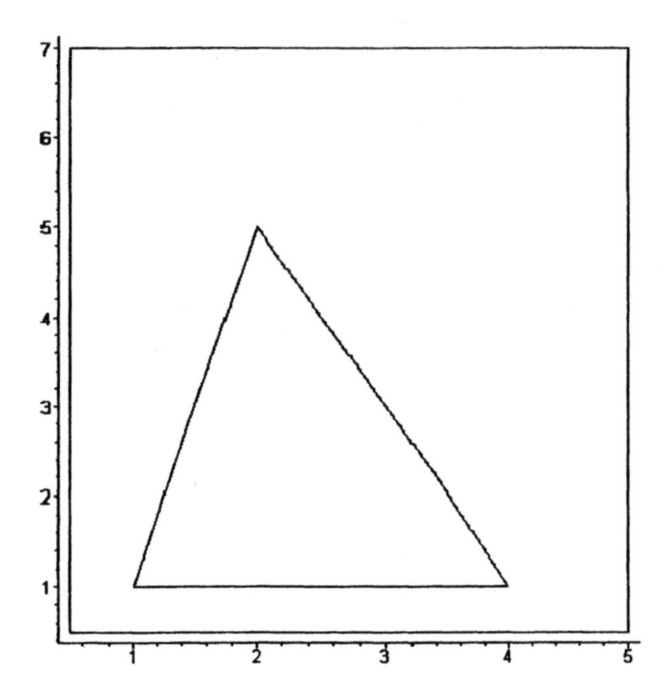

Notice the use of the colon after the two listplots to suppress unwanted output. Also note we have given each listplot its own name.

Now for plotting curves such as circles. Recall from Calculus that a circle, centered at the origin, of radius R is give parametrically as $(R\cos(\theta), R\sin(\theta))$. Lets see what happens to a circle of radius 2, centered at the origin, when it is mapped by the matrix transformation $A = \begin{bmatrix} 1 & 2 \\ -3 & 5 \end{bmatrix}$. Setting $x = (2\cos(\theta), 2\sin(\theta))$ plot the circle x and also its image under the matrix A that is Ax.

```
>plot([2*cos(t),2*sin(t),t=0..2*Pi],x=-2..2);
```

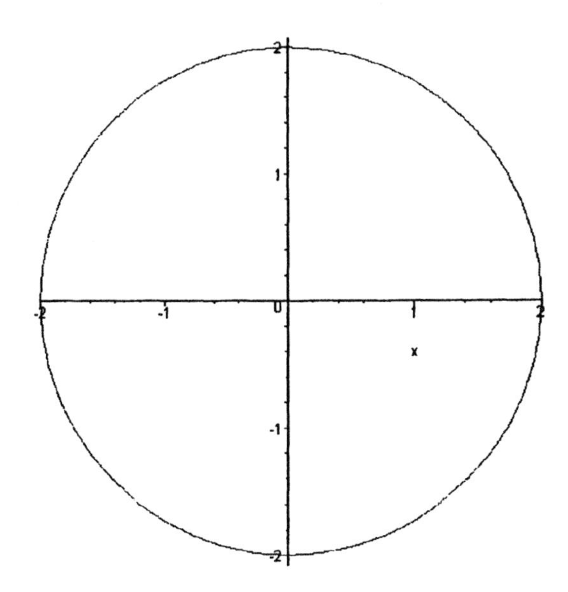

The image Ax is $Ax = (2\cos(\theta) + 4\sin(\theta), -6\cos(\theta) + 10\sin(\theta))$. The plot Ax is

```
> plot([2*cos(t)+4*sin(t),-6*cos(t)+10*sin(t),t=0..2*Pi]);
```

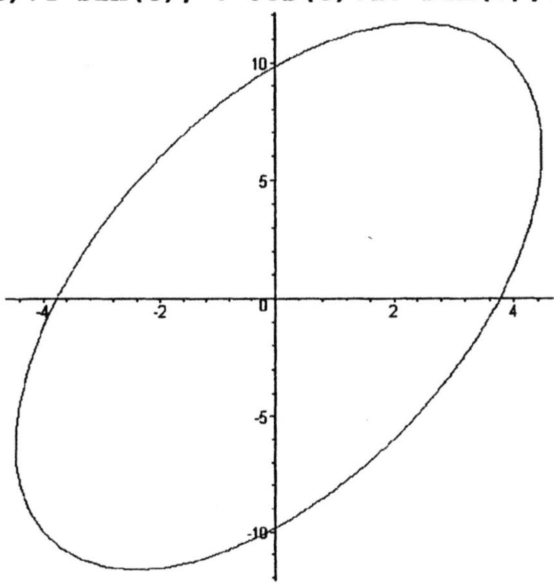

We see that the image is an ellipse whose major axes has been rotated counterclockwise. You will be learning about matrix transformations in your study of linear algebra. Right now we will just take for granted that the transformation given above is correct.

Exercises

1). Plot each of the following functions choosing a different color for each plot.

a). y = sin(2x)cos(3x).

b). y = ln(5x).

c). $y = 3e^x$.

2). Plot the following groups of functions on the same graph.

a). y = sin(2x), y = $x^2 + 4x - 5$.

b). y = ln(5x), y = e^x, y = 3ln(x) + $2e^x$.

3). Plot the diamond shaped figure with vertices at (0,1), (1,0), (0,-1), (-1,0).

4). Plot the diamond in problem (3) after shifting its center to the coordinates (-3,5).

5). Consider the square with coordinates (0,0), (2,0), (2,2), (0,2). Shift the center of the square and the diamond in problem (3) to the coordinates (1,1) and plot them together in a single plot.

Chapter 2.0

Basic operations using the linalg package

In this chapter we will be using the commands and syntax associated with the **linalg** package. We will be working with both vectors and matrices. To introduce matrices consider the system of equations

$$a_{11}x_1 + a_{12}x_2 + a_{13}x_3 = b_1$$
$$a_{21}x_1 + a_{22}x_2 + a_{23}x_3 = b_2 .$$
$$a_{31}x_1 + a_{32}x_2 + a_{33}x_3 = b_3$$

To solve this set of equations note that the x's serve merely as place holders and all we really need to pay attention to are the a's and the b's. This is most easily organized by putting the a's into a matrix and the b's into a vector as follows.

$$A = \begin{bmatrix} a_{11} & a_{12} & a_{13} \\ a_{21} & a_{22} & a_{23} \\ a_{31} & a_{32} & a_{33} \end{bmatrix} \text{and}\, b = \left[b_1, b_2, b_3\right]^T .$$

The first column of A contains the coefficient multipliers of x_1. Similarly the second column of the matrix A contains the coefficient multipliers of x_2 and so on. The superscript of T on the vector b indicates that it is a column vector and not a row vector. T stands for transpose, a matrix operation that will be demonstrated below. In connection with matrices we shall use the general notation $A = \left[a_{ij}\right]$ for a general matrix and if we wish to emphasize the fact that it has m rows and n columns then we will write $A = \left[a_{ij}\right]_{m\times n}$. The system of equations can now be written as the matrix equation Ax = b where $x = \left[x_1, x_2, x_3\right]^T$.

Our purpose is to show how to create matrices and vectors in maple using the **linalg** package. We will also see how to implement the commonly used matrix and vector operations.

1. Basic Matrix Operations

The basic operations are those of multiplication of a matrix by a constant, adding two matrices and multiplying matrices together. We begin by defining several matrices. The maple commands are shown in red while maple's output is shown in blue.

```
> A:=matrix([[1,2],[-3,1]]); B:=matrix([[4,-5],[5,7]]);
> C:=matrix([[2,3],[-4,1],[2,-5]]);
```

$$A := \begin{bmatrix} 1 & 2 \\ -3 & 1 \end{bmatrix}$$

$$B := \begin{bmatrix} 4 & -5 \\ 5 & 7 \end{bmatrix}$$

$$C := \begin{bmatrix} 2 & 3 \\ -4 & 1 \\ 2 & -5 \end{bmatrix}$$

```
> D1:=matrix([[1,-2,3],[3,-2,4]]);
```

$$D1 := \begin{bmatrix} 1 & -2 & 3 \\ 3 & -2 & 4 \end{bmatrix}$$

Note that we enter matrices by rows. The name D1 is used in place of D because D is a reserved word in maple. The use of D would generate an error message. Now for some elementary matrix operations. The maple function **evalm** is used to reduce matrix/vector operations down to a single vector or matrix which ever is appropriate. It will be liberally used in the examples below.

```
> A1:=evalm(5*A); # multiply A by the constant 5.
```

$$A1 := \begin{bmatrix} 5 & 10 \\ -15 & 5 \end{bmatrix}$$

```
> A2:=evalm(A+B); # add A and B together.
```

$$A2 := \begin{bmatrix} 5 & -3 \\ 2 & 8 \end{bmatrix}$$

```
> A3:=evalm(A&*D1); # the matrices must be of appropriate
                    # sizes.
```

$$A3 := \begin{bmatrix} 7 & -6 & 11 \\ 0 & 4 & -5 \end{bmatrix}$$

Note that when multiplying A and D1 together that the operator is &*. This is the operator for matrix multiplication. Note also that when computing A1 we used * as the operator for multiplication by a constant. The use of **evalm** and the correct operator is crucial to success. The transpose operation was introduced earlier. As an example we take the transpose of C.

```
> C1:=evalm(transpose(C));   # We write C1 =
```
C^T

$$C1 := \begin{bmatrix} 2 & -4 & 2 \\ 3 & 1 & -5 \end{bmatrix}$$

Note that the first column of C is now the first row of C1 while the second column of C is now the second row of C1. For some linear combinations consider:

```
> A1:=evalm(5*A-7*B);
```

$$A1 := \begin{bmatrix} -23 & 45 \\ -50 & -44 \end{bmatrix}$$

```
> A2:=evalm(17*transpose(C)-15*D1);
```

$$A2 := \begin{bmatrix} 19 & -38 & -11 \\ 6 & 47 & -145 \end{bmatrix}$$

```
> A3:=evalm(B&*(17*transpose(C)-15*D1));
```

$$A3 := \begin{bmatrix} 46 & -387 & 681 \\ 137 & 139 & -1070 \end{bmatrix}$$

Now for some vector operations. We can multiply a vector by a constant and add two vectors together which gives us the ability to form linear combinations. For example

```
> x:=vector([1,2,3,4]);  y:=vector([3,-2,-1,4]);
```

$$x := [1, 2, 3, 4]$$

$$y := [3, -2, -1, 4]$$

```
> x1:=evalm(5*x); # multiply the vector x by 5
```

$$x1 := [5, 10, 15, 20]$$

```
> x2:=evalm(x+y); # add the vectors x and y together
```

$$x2 := [4, 0, 2, 8]$$

```
> x3:=evalm(4*x-17*y); # Form the linear combination 4x-17y
```

$$x3 := [-47, 42, 29, -52]$$

2. Important Vector Concepts

There are additional vector operations. Let $x = [x_1, x_2, \ldots, x_n]^T$ where again T means transpose from a row vector to a column vector. The vector x has n components with x_1 being the first component and x_n being the nth component. One important property of a vector is its length which is defined to be $\|x\| = \left\{ \sum_{k=1}^{n} x_k^2 \right\}^{1/2}$. This can be computed using the **norm** function. If the length is one then the vector is said to be a unit vector. Another important operation is that of the innerproduct of two vectors usually denoted by $x \circ y$. If $y = [y_1, y_2, \ldots, y_n]^T$ then $x \circ y = \sum_{k=1}^{n} x_k y_k$, For the case of $n \le 3$ we have

$x \circ y = \sum_{k=1}^{n} x_k y_k = \|x\| \|y\| \cos\theta$ where θ is the angle between the two vectors x and y. This enables us to get at the angle between the two vectors. The vectors x and y are said to be orthogonal (that is perpendicular) if $x \circ y = 0$ For the case of $n \le 3$ this means that the angle θ between the vectors x and y is $\frac{\pi}{2}$. Two vectors are parallel, that is have the same or exactly opposite directions, if there is a constant λ such that x = λy. The following examples illustrate these ideas. The very first operation is to read in the linear algebra package and note that the command is terminated with a colon to suppress unwanted output.

```
> with(linalg):
Warning, the protected names norm and trace have been redefined and
unprotected
```

```
>x:=vector([1,2,3,4]); y:=vector([-2,-3,1,5]);
```

$$x := [1, 2, 3, 4]$$

$$y := [-2, -3, 1, 5]$$

```
>a1:=dotprod(x,y);   # The innerproduct of x and y is 15
```

$$a1 := 15$$

```
> a2:=norm(x,2);   a3:=norm(y,2); # norm(x,2) gives the
                                  # length of x
```

$$a2 := \sqrt{30}$$

$$a3 := \sqrt{39}$$

Example 2.1

Determine if the vectors $x = [1,-1,2]$, $y = [1,1,0]$, $u = [4,4,-8]$ are orthogonal, parallel, or neither.

Solution

```
> x:=vector([1,-1,2]); y:=vector([1,1,0]);
> u:=vector([4,4,-8]); zero:=vector(3,0);
```

$$x := [1, -1, 2]$$

$$y := [1, 1, 0]$$

$$u := [4, 4, -8]$$

$$zero := [0, 0, 0]$$

```
>a1:=dotprod(x,y);    # x and y are othogonal, the
                        # innerproduct is zero
```

$$a1 := 0$$

```
>a2:=dotprod(x,u);    # x and u are not othogonal, the
                        # innerproduct is not zero
```

$$a2 := -16$$

```
>a3:=dotprod(y,u);    # y and u are not othogonal, the
                        # innerproduct is not zero
```

$$a3 := 8$$

```
>y1:=evalm(lambda*y);
```

$$y1 := [\lambda, \lambda, 0]$$

Now we shall test to see if x and y are parallel by seeing if there is a λ that will make x - y1 = 0.

```
>w1:=evalm(x-y1)=evalm(zero); # Note that w1 is an equation
```

$$w1 := [1 - \lambda, -1 - \lambda, 2] = [0, 0, 0]$$

They are clearly not parallel for no choice will make the third component, 2, zero. Using the same technique on the other vectors we have

```
> u1:=evalm(lambda*u);
```

$$u1 := [4\lambda, 4\lambda, -8\lambda]$$

```
> w2:=evalm(x-u1)=evalm(zero);
```

$$w2 := [1-4\lambda, -1-4\lambda, 2+8\lambda] = [0, 0, 0]$$

x and u are not parallel for $\lambda = 1/4$ is required to make the first component 0 while $\lambda = -1/4$ is required to make the third component 0. Finally we check y and u.

```
> w3:=evalm(y-u1)=evalm(zero);
```

$$w3 := [1-4\lambda, 1-4\lambda, 8\lambda] = [0, 0, 0]$$

$\lambda = 0$ will make the third component 0 but will not make the first or second component 0 hence y and u are not parallel.

Next we consider operations that involve both matrices and vectors. Let

```
> A:=matrix([[1,2],[3,4],[5,6]]);
> B:=matrix([[-2,1],[3,-5]]);C:=matrix([[4,-1],[7,-8]]);
```

$$A := \begin{bmatrix} 1 & 2 \\ 3 & 4 \\ 5 & 6 \end{bmatrix}$$

$$B := \begin{bmatrix} -2 & 1 \\ 3 & -5 \end{bmatrix}$$

$$C := \begin{bmatrix} 4 & -1 \\ 7 & -8 \end{bmatrix}$$

```
> F:=matrix([[1,-2,0],[4,0,-7]]); x:=vector([1,-3,2]);
> y:=vector([7,0,-2]); u:=vector([6,3,-1]);
```

$$F := \begin{bmatrix} 1 & -2 & 0 \\ 4 & 0 & -7 \end{bmatrix}$$

$$x := [1, -3, 2]$$

$$y := [7, 0, -2]$$

$$u := [6, 3, -1]$$

Example 2.2

Determine the value of the linear combination A(3B-5C)F(16x-18y+3u).

Solution

```
> w:=evalm(16*x-18*y+3*u);
```

$$w := [-92, -39, 65]$$

```
> G:=evalm(A&*(3*B-5*C)&*F);
```

$$G := \begin{bmatrix} 154 & 156 & -406 \\ 314 & 364 & -868 \\ 474 & 572 & -1330 \end{bmatrix}$$

```
> answer:=evalm(G&*w);   # Note the use of the operator &*.
```

$$answer := [-46642, -99504, -152366]$$

Example 2.3

For what values of p and q , if any, will A = B where $A = \begin{bmatrix} 2 & 3p-4q \\ 1 & 5 \end{bmatrix}$

and $B = \begin{bmatrix} 2 & 4p+5q \\ q & 5 \end{bmatrix}$?

Solution

```
> A:=matrix([[2,3*p-4*q],[1,5]]);
> B:=matrix([[2,4*p+5*q],[q,5]]);
```

$$A := \begin{bmatrix} 2 & 3\,p-4\,q \\ 1 & 5 \end{bmatrix}$$

$$B := \begin{bmatrix} 2 & 4\,p+5\,q \\ q & 5 \end{bmatrix}$$

```
> C:=evalm(A-B);
```

$$C := \begin{bmatrix} 0 & -p-9\,q \\ 1-q & 0 \end{bmatrix}$$

Clearly the entry in row 2 and column 1 (position [2,1]) requires q = 1. Armed with this then the entry [1,2] requires -p - 9 = 0 or p = -9. Thus there is only one set of values that will make the two matrices equal.

3. Vector Projection

In physics and engineering forces are frequently used. In many cases the given force must be broken up into the sum of two mutually perpendicular forces whose sum is the given vector. Furthermore one of the two perpendicular forces must have a prescribed direction that is be parallel to a given vector. This problem is most easily solved by looking at the problem of finding the projected image of one vector, say a, onto a specified vector say b. This is illustrated in the figure below.

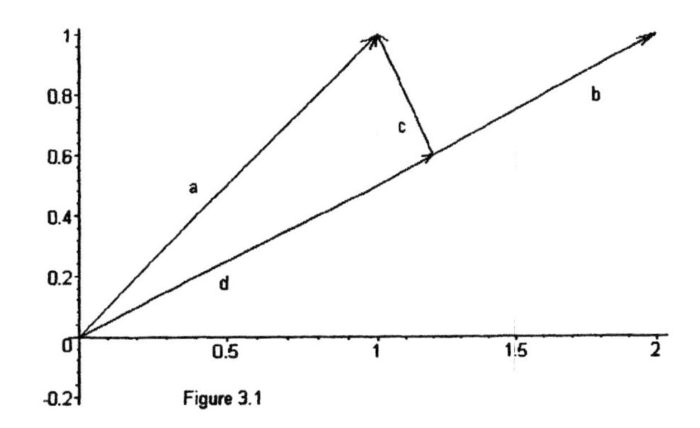

Figure 3.1

The vector a is to be represented as a = d + c where d is parallel to the specified direction b and c is perpendicular to d. To solve this problem let θ be the angle between the vectors a and b we then have

$$\begin{aligned}\|d\| &= \|a\|\cos(\theta)\\ &= \|a\|\frac{a \circ b}{\|a\|\|b\|}\\ &= \frac{a \circ b}{\|b\|}\end{aligned}$$

This last result is the length of the vector d. All that we need to do now is to create a unit vector in the direction of b and multiply it by the length of the vector d as in $d = \frac{a \circ b}{\|b\|}\frac{b}{\|b\|} = \left(\frac{a \circ b}{b \circ b}\right)b$. The vector c can now be found as c = a – d.

Example 3.1

Given the vectors $x = [4,-5,7]^T$ and $y = [-8,-17,2]^T$ represent x and the sum of two mutually perpendicular vectors one of which is parallel to y.

Solution

```
> restart; with(linalg):
Warning, the protected names norm and trace have been redefined and
unprotected

> x:=vector([4,-5,7]);     y:=vector([-8,-17,2]);
```

$$x := [4, -5, 7]$$

$$y := [-8, -17, 2]$$

```
> a1:=dotprod(x,y);     a2:=dotprod(y,y);
```

$$a1 := 67$$

$$a2 := 357$$

```
> z1:=evalm((a1/a2)*y); z2:=evalm(x-z1); # x = z1 + z2
```

$$z1 := \left[\frac{-536}{357}, \frac{-67}{21}, \frac{134}{357}\right]$$

$$z2 := \left[\frac{1964}{357}, \frac{-38}{21}, \frac{2365}{357}\right]$$

Now z1 and z2 are perpendicular by construction. Just as an exercise we shall verify this. We have

```
> w:=dotprod(z1,z2);
```

$$w := 0$$

Example 3.2

In the figure below the box is being pushed up the ramp with the given force. The ramp has a ten percent grade, that is for every foot forward the ramp rises 1/10 th of a foot. The given force is F = [300,-50]. Break this force up into the sum of two forces F = A + B where A is parallel to the ramp and B is perpendicular to the ramp.

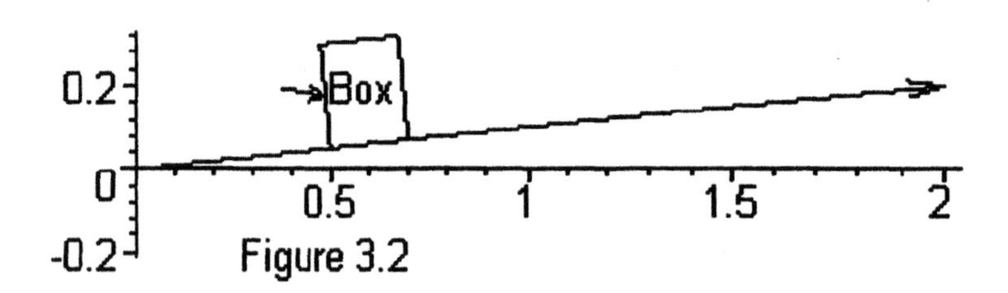

Figure 3.2

Solution

A vector in the direction of the ramp is R = [1,0.1]. The vector A is then the projection of F onto R. We have

```
> R:=vector([1,0.1]); F:=vector([300,-50]);
```

$$R := [1, 0.1]$$

$$F := [300, -50]$$

```
> a1:=dotprod(F,R);   a2:=dotprod(R,R);
```

$$a1 := 295.0$$

$$a2 := 1.01$$

```
> A:=evalm(a1*R/a2);
```

$$A := [292.0792079, 29.20792079]$$

```
> B:=evalm(F-A);
```

$$B := [7.9207921, -79.20792079]$$

4. Graphs and Incidence Matrices

For another application of matrices we turn to graph theory. A graph is a set of nodes and edges which connect some of the nodes with a path (here a line segment). Consider the graph below in figure 3.

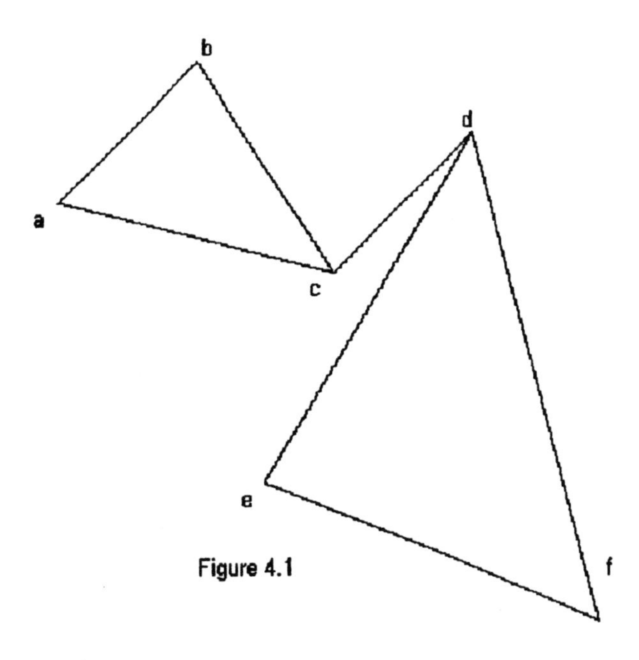

Figure 4.1

We can represent this graph by means of its incidence matrix A. In the [i,j] entry of the incidence matrix we place a one if there is a path (line segment) joining nodes i and j otherwise we set it equal to 0. The incidence matrix is then

$$A = \begin{array}{c} \begin{matrix} a & b & c & d & e & f \end{matrix} \\ \begin{bmatrix} 0 & 1 & 1 & 0 & 0 & 0 \\ 1 & 0 & 1 & 0 & 0 & 0 \\ 1 & 1 & 0 & 1 & 0 & 0 \\ 0 & 0 & 1 & 0 & 1 & 1 \\ 0 & 0 & 0 & 1 & 0 & 1 \\ 0 & 0 & 0 & 1 & 1 & 0 \end{bmatrix} \end{array} \begin{matrix} a \\ b \\ c \\ d \\ e \\ f \end{matrix}$$

Here we have identified a with 1, b with 2, etc. Thus A[3,4] is the connection, if any, between nodes c and d. Since there is a path between them A[3,4] = 1. As we see the incidence matrix contains exactly the same information as does the original graph. How can we take advantage of this? To make use of the incidence matrix we need to recall how matrix multiplication is done. If

$$A=\left[a_{ij}\right]_{m\times n} \text{ and } B=\left[b_{ij}\right]_{n\times r} \text{ then } AB=\left[\sum_{k=1}^{n}a_{ik}b_{kj}\right]_{m\times r}. \text{ Now}$$

AA[i,j]= $a_{i1}a_{1j}+a_{i2}a_{2j}+a_{i3}a_{3j}+\cdots+a_{in}a_{jn}$. If $a_{ik}a_{kj}$=1 then there is a path from node i to node k and a path from node k to node j thus there is a path of length 2 from node i to node j. AA[i,j] is the total number of paths of length 2 from node i to node j. Similarly AAA[i,j] is the total number of paths of length 3 from node i to node j. In general $A^k[i,j]$ is the total number of paths of length k from node i to node j. Let us compute some powers of the incidence matrix for the above graph.

```
>A:=matrix([[0,1,1,0,0,0],[1,0,1,0,0,0],[1,1,0,1,0,0],[0,0,1
,0,1,1],[0,0,0,1,0,1],[0,0,0,1,1,0]]);
```

$$A:=\begin{bmatrix}0&1&1&0&0&0\\1&0&1&0&0&0\\1&1&0&1&0&0\\0&0&1&0&1&1\\0&0&0&1&0&1\\0&0&0&1&1&0\end{bmatrix}$$

```
> A2:=evalm(A^2);
```

$$A2:=\begin{bmatrix}2&1&1&1&0&0\\1&2&1&1&0&0\\1&1&3&0&1&1\\1&1&0&3&1&1\\0&0&1&1&2&1\\0&0&1&1&1&2\end{bmatrix}$$

AA[3,3] = 3 tells us that there are 3 paths of length 2 joining nodes c to c. They are c-a-c, c-b-c, and c-d-c. AA[3,5] = 1 tells us that there is one path of length 2 joining nodes c and e it is c-d-e.

```
> A4:=evalm(A^4);
```

$$A4:=\begin{bmatrix}7&6&6&6&2&2\\6&7&6&6&2&2\\6&6&13&4&6&6\\6&6&4&13&6&6\\2&2&6&6&7&6\\2&2&6&6&6&7\end{bmatrix}$$

AAAA[6,1] = 2 tells us that there are 2 paths of length 4 joining nodes f and a, they are a-c-d-e-f and a-b-c-d-f.

5. Markov Chains

In our next example we shall look at an example of a Markov chain. A Markov matrix is one in which every entry satisfies $0 \le a_{ij} \le 1$ and where the sum of the entries in each column is one. For this example we shall set C = cloudy, R = rainy, and S = sunny. We assume that C, R, and S are the only weather possibilities. Set

$$\begin{array}{c} \text{today} \\ \begin{array}{ccc} S & C & R \end{array} \end{array}$$

$$A=\begin{bmatrix} \frac{5}{8} & \frac{3}{8} & \frac{5}{8} \\ \frac{1}{4} & \frac{3}{8} & \frac{1}{4} \\ \frac{1}{8} & \frac{1}{4} & \frac{1}{8} \end{bmatrix} \begin{array}{l} S \\ C \\ R \end{array} \text{ tomorrow}$$

We interpret the matrix A as follows. If it is sunny today then the probability that it will be cloudy tomorrow is A[2,1] = 1/4 or 0.25. If it is cloudy today then the probability that it will be sunny tomorrow is A[1,2] = 3/8.We enter the data into the matrix in rational form because this will give more accurate results when we compute powers of A. If we had entered the results in decimal form we would have gotten less accurate results when computing powers of A.

Now let S0 = [1/2,1/4,1/4] be the probabilities of S, C, and R for today's weather. The probability for tomorrows weather then is S1 = A(S0). The probability for the weather the day after tomorrow will be S2 = A(S1) = AA(S0) etc. Computing these quantities gives .

```
> A:=matrix([[5/8,3/8,5/8],[1/4,3/8,1/4],[1/8,1/4,1/8]]);
```

$$A := \begin{bmatrix} \frac{5}{8} & \frac{3}{8} & \frac{5}{8} \\ \frac{1}{4} & \frac{3}{8} & \frac{1}{4} \\ \frac{1}{8} & \frac{1}{4} & \frac{1}{8} \end{bmatrix}$$

```
> S0:=vector([1/2,1/4,1/4]);
```

$$S0 := \left[\frac{1}{2}, \frac{1}{4}, \frac{1}{4}\right]$$

```
> S1:=evalm(A&*S0);
```

$$S1 := \left[\frac{9}{16}, \frac{9}{32}, \frac{5}{32}\right]$$

```
> S2:=evalm(A&*S1);
```

$$S2 := \left[\frac{71}{128}, \frac{73}{256}, \frac{41}{256}\right]$$

Now lets see if there is a long term trend. Lets compute the 100th and 101st power of A and apply it to S0 to see if the sequence of S's is converging to anything. We are also going to want to see the resulting entries in the vector in decimal form. To convert we will use the **map** function. The symbol \ is used to denote the continuation of a number in maple.

```
> S100:=evalm(A^100&*S0); S100A:=map(evalf,S100);
```

S100 := [2255289830941752452654350583596097249735554292987286563204298 \
  3546423507247379799247030470095 / 40740719526689721725368913768 18 \
  75632210293678733187250127228089870876259952667341236679 4752 , 23280\
  41115810841241449652215325003612630249592761070000727017656405007 \
  19972952766420959 7001 / 81481439053379443450737827536375126442 05 \
  873574663745002544561797417525199053346824733589504 , 13095231276435\
  98198315429371120314532104515395928101875408947431727816549847859 \
  311117898313 / 8148143905337944345073782753637512644205873574663 \
  745002544561797417525199053346824733589504]

S100A := [0.5535714286, 0.2857142857, 0.1607142857]

Note that the entries in the vector S100 are in rational form and hence are more accurate than their ten digit approximations in the vector S100A.

```
> S101:=evalm(A&*S100); S101A:=map(evalf,S101);
```

S101 := [18042318647534019621234804668768777997884434343898292505634 38 \
  6837138805797903839397624376759 / 325925756213517773802951310145 \
  5005057682349429865498001017824718967010079621338729893435 8016 , 186\
  2432892648672993159721772260002890104199674208856000581614125 1240 \
  0575978362213136767760 09 / 65185151242703554760590262029100101 15 \
  36469885973099600203564943793402015924267745978687160 32 , 1047618502\
  11487855865234349689625162568361231674248150032715794538225323987 \
  82874488943186505 / 6518515124270355476059026202910010115364698 8 \
  597309960020356494379340201592426774597868716032]

S101A := [0.5535714286, 0.2857142857, 0.1607142857]

To see how close together these last two vectors are, we shall look at the difference w:=S101 - S100

```
> w:=evalm(S101-S100); w1:=map(evalf,w); w2:=evalm(S101A-
S100A);
```

w := [-1/325925756213517773802951310145500505768234942986549800101782 4 \
718967010079621338729893435801 6, 1/65185151242703554760590262029 10 \
010115364698859730996002035649437934020159242677459786871603 2 , 1/6\
51851512427035547605902620291001011536469885973099600203564943793 \
402015924267745978687160 32]

$w1 := [-0.3068183416\ 10^{-91}, 0.1534091708\ 10^{-91}, 0.1534091708\ 10^{-91}]$

$$w2 := [0., 0., 0.]$$

This result shows that the vectors are nearly equal and so the sequence of S's seems to be converging. Note that the difference in the vectors S100 and S101 does not result in the zero vector while subtracting S101A from S100A does result in the zero vector. This shows that the rational entries are more accurate than the ten digit floating point entries.

6. Complex Values

Maple can handle complex as well as real values in both vectors and matrices. We enter complex values as 3 + 4*I where I stands for the square root of -1. For some examples we have:

```
> restart; with(linalg):
Warning, the protected names norm and trace have been redefined and
unprotected
```

```
> a:=2+3*I;   b:=4-5*I;
```

$$a := 2 + 3I$$

$$b := 4 - 5I$$

```
> w:=a*b;
```

$$w := 23 + 2I$$

```
> a1:=conjugate(a); w1:=a1*b;
```

$$a1 := 2 - 3I$$

$$w1 := -7 - 22I$$

For complex valued vectors the innerproduct of two vectors x and y is defined to be $x \circ y = \sum_{k=1}^{n} \bar{x}_k y_k$ where the horizontal bar over x_k indicates the conjugate of x_k. This definition reduces to the one given earlier for real vectors. The maple function **dotprod** applies this definition and so the innerproduct of complex vectors will be computed correctly. For some examples we have:

```
> A:=matrix([[2+3*I,4-5*I],[6+7*I,1-2*I]]);
```

$$A := \begin{bmatrix} 2+3I & 4-5I \\ 6+7I & 1-2I \end{bmatrix}$$

```
> x:=vector([1-7*I,13+21*I]); y:=vector([2-5*I,17+15*I]);
```

$$x := [1 - 7\,I, 13 + 21\,I]$$

$$y := [2 - 5\,I, 17 + 15\,I]$$

```
> B:=evalm(A&*x);
```

$$B := [180 + 8\,I, 110 - 40\,I]$$

```
> w:=dotprod(x,y);
```

$$w := 573 + 153\,I$$

Exercises

Given the matrices;

$$A = \begin{bmatrix} -4 & 2 \\ 5 & -6 \end{bmatrix},\quad B = \begin{bmatrix} 7 & -3 \\ 2 & 10 \end{bmatrix},\quad C = \begin{bmatrix} 15 & 12 \\ -18 & 26 \end{bmatrix},$$

$$F = \begin{bmatrix} 3 & 0 & 17 & 21 \\ 0 & -7 & 9 & -23 \end{bmatrix},\quad G = \begin{bmatrix} 1 & 3 \\ 2 & 7 \\ -5 & 12 \\ 0 & 18 \end{bmatrix}$$ and the vectors

$$x = [2,-3]^T,\quad y = [-5,8]^T,\quad z = [12,21]^T$$
$$u = [2,5,-8,17]^T,\quad v = [-18,21,0,-6]^T.$$
Solve the following problems:

1). Find Fv.
2). Find Ax.
3). Find the linear combination 3x - 17y + 21z.
4). Find the linear combination 5A – 8B + 25C.
5). Find the matrix products AG^T.
6). Find the matrix product FG.
7). Find the matrix/vector product Gx.
8). Find the matrix /vector product Fu.
9). Find the linear combination G(Fv + C(5A – 127B)(17x – 32y + 25z)).
10). Write the vector u as the sum of two mutually perpendicular vectors one of which is parallel to v.
11). Write the vector v as the sum of two mutually perpendicular vectors one of which is parallel to u.
12). For the graph shown above determine the total number of paths of length ≤ 6 between any two nodes.
13). For the graph shown above determine how many paths of length 10 exist between nodes a and f.
14). For the graph shown above determine how many paths of length 15 exist between nodes b and e.
15). For the Markov weather matrix determine the terms in the sequence S1, S2, …, S10 starting with the given S0
16). Starting with S0 generate the terms, with rational entries, in the sequence S1, S2, …, Sn, … In each case compute the differences S2-S1, S3-S2, etc until all the entries in

the difference vector are $\le 10^{-4}$. What is the smallest value of n that suffices?

Project 1.

Consider a very small store that sells only five items, weed eaters, garden hoses, rakes, hoes, and twig cutters. They carry only one brand and model of each item. When their inventory grows low they reorder the item to bring their inventory up to maximum level.

The owner is interested in several items of information. For instance after this months sales are taken into account what items need to be reordered. For those being reordered how many should be ordered. To answer questions of this nature we can use vectors and vector operations.

We shall express all items as a vector of length 5 where the items in the vector refer to [weed eater, garden hose, rake, hoe, twig cutter].

To answer the questions below we need the following information.

Inventory prior to this months sales	[50,80,20,20,40].
Units sold this month	[35,45,5,6,28].
Reorder if the current inventory equals or falls below	[20,35,5,5,20].
Customer Purchase Price per unit in dollars	[95,25,15,18,22].
Reorder Price per unit in dollars	[65,15,8,10,17].
Reorder to the following level of units	[70,100,25,25,50].

Using vectors answer the following questions:

1). Find the current inventory, that is the inventory after this months activity.

2). From the answer to problem (1) which items, if any, need to be reordered?

3). From (2) how many of each item needs to be reordered?

4). What is the total cost of the reordered items?

5). What is the total value of this months sales?

6) What is the cost to the owner of all the items in the new inventory after the ordered items have been received?

Project 2

Consider a very small store that sells only five items, weed eaters, garden hoses, rakes, hoes, and twig cutters. We shall keep information on these items in both matrix and vector form. When they are kept in vector form the items will be ordered as follows [w,g,r,h,t]. Where w = weed eaters, g = garden hose, r = rake, h = hoes, and t = twig cutters. For some data we have

Customer Purchase Price per unit in dollars [95,25,15,18,22].
Reorder Price per unit in dollars [65,15,8,10,17].

Units sold in the given month

$$A = \begin{array}{c} \begin{matrix} w & g & r & h & t \end{matrix} \\ \begin{bmatrix} 32 & 12 & 8 & 4 & 13 \\ 41 & 24 & 13 & 11 & 21 \\ 53 & 32 & 18 & 19 & 28 \\ 62 & 28 & 15 & 17 & 35 \\ 27 & 25 & 29 & 8 & 33 \\ 11 & 15 & 37 & 3 & 23 \end{bmatrix} \end{array} \begin{array}{r} \\ \text{April} \\ \text{May} \\ \text{June} \\ \text{July} \\ \text{August} \\ \text{September} \end{array}$$

To interpret this matrix we note that A[3,3] = 18 is the number of rakes sold in June while A[5,6] = 33 is the number of twig cutters sold in the month of August. You will need to use this matrix along with the price vector to answer the following questions.

Hint: Note that if x = [1,1,1,1,1] and w = xA then w1 the first element of w is the sum of all of the elements in column 1 of A. w2 the second element of w is the sum of all of the elements in column 2 of A etc. See the maple commands **row** and **column.**

Answer the following questions:

1). How many units of each item were sold in the 6 months?

2). What is the revenue for the sale of the items in the answer to problem (1)?

3). What is the total value of the sale of rakes in the three months of June, July, and August.

4). What is the total value of all of the rake sales in the 6 month period?

For the following problems we will need

Starting inventory [50,80,20,20,40].

Reorder if the current inventory falls below [20,35,30,20,20].

Reorder to the following level of units [70,100,25,25,50].

Thus at any time the current inventory (computed after the end of each month) equals or falls below the reorder minimum we must reorder to bring the inventory up to the specified limit. For instance if at the end of some month the number of garden hoses currently on hand is 22 then we must reorder 100 – 22 = 78 hoses to bring the inventory up to the maximum of 100 hoses.

5). Determine the first month when each item must be reordered. Note that his will be different for different items,

6). For weed eaters determine the specific months in which this item must be reordered and also the total number of weed eaters reordered over the 6 month period.

Hint: For questions 5 and 6 you may want to use loops along with the **if-then-else** statement.

Project 3

A large object is being moved along an inclined plane which has a 15% grade. That is for each foot forward the inclined place rises 0.15 foot. The object is acted on by 4 forces the x (column 1) and y (colu,m 2) components of which we keep in the matrix A. Answer the following questions, where the matrix containing the 4 forces is

$$A = \begin{bmatrix} 100 & -20 \\ 50 & 15 \\ -40 & -30 \\ -15 & 5 \end{bmatrix}$$

Hint: See the hint for project number 2.

1). The resultant force R is the sum of the 4 forces. Find R as a vector with components [x,y].

2). Express R as R = R1 + R2 where R1 is parallel to the inclined plane and R2 is perpendicular to R1.

3). Express A as A = A1 + A2 where the vector in the first row of A is the sum of the vectors in the first row of A1 and A2. Furthermore row(A1,i) is parallel to the inclined plane and row(A2,i) is perpendicular to row(A1,i) for i = 1, 2, 3, 4.

4). Add up the columns of A1 and A2 in problem (3).

5). Compare the vectors from problems (4) and (2).

6). Is the object being moved up or down the plane?

Project 4

Consider the graph below

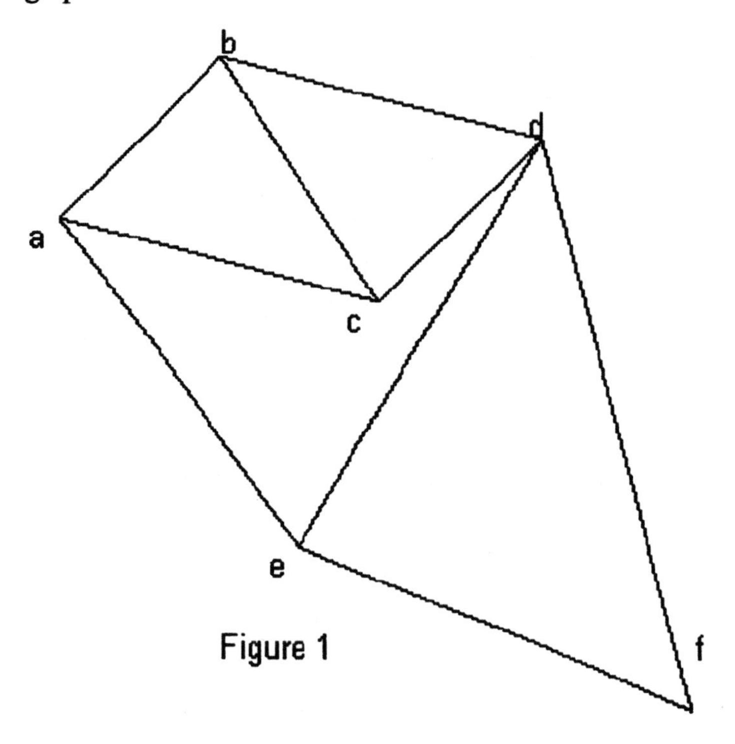

Figure 1

For the graph in Figure 1 answer the following questions:

1). What is the incidence matrix A? Create the matrix for later use.

2). Find all paths of length 2.

3). For the distinct paths of length two A[3,2], A[2,3] and A[4,3] give the node connections a-e-d etc.

4). Find all paths of length three. For the distinct paths of A[3,2] show the node connections such as a-e-f-d etc.

5). Find all paths of length 4. For A[3,2] find 4 of the distinct paths and show the node connections such as a-e-f-d-c etc.

6). Find all paths of length less than or equal to 8. You may want to use the **sum** function.

Chapter 2.1
Basic operations using the Linear Algebra package

In this chapter we will be using the commands and syntax associated with the **LinearAlgebra** package. This is the package that we would use to solve real world computational problems. The algorithms it uses are designed to produce highly accurate solutions. For real world large problems it would be the preferred package to use for solving systems of equations. We will be working with both vectors and matrices. To introduce matrices consider the system of equations

$$a_{11}x_1 + a_{12}x_2 + a_{13}x_3 = b_1$$
$$a_{21}x_1 + a_{22}x_2 + a_{23}x_3 = b_2 \ .$$
$$a_{31}x_1 + a_{32}x_2 + a_{33}x_3 = b_3$$

To solve this set of equations note that the x's serve merely as place holders and all we really need to pay attention to are the a's and the b's. This is most easily organized by putting the a's into a matrix and the b's into a vector as follows.

$$A = \begin{bmatrix} a_{11} & a_{12} & a_{13} \\ a_{21} & a_{22} & a_{23} \\ a_{31} & a_{32} & a_{33} \end{bmatrix} \text{and}\, b = [b_1, b_2, b_3]^T .$$

The superscript of T on the vector b indicates that it is a column vector and not a row vector. T stands for transpose a matrix operation that will be demonstrated below. In connection with matrices we shall use the general notation $A = \left[a_{ij}\right]$ for a general matrix and if we wish to emphasize the fact that it has m rows and n columns then we will write $A = \left[a_{ij}\right]_{m \times n}$. The system of equations can now be written as the matrix equation Ax = b where $x = [x_1, x_2, x_3]^T$.

Our purpose is to show how to create matrices and vectors in maple using the **LinearAlgebra** package. We will also show how to implement the commonly used matrix and vector operations.

1. Basic Matrix Operations

We shall demonstrate two ways of declaring matrices and vectors. The commands in the **LinearAlgebra** package generally start with a capital letter. Since maple is sensitive to the case of the letters we need to exercise caution in our typing. The first step is to read in the **LinearAlgebra** package, and then take a look at some examples.

```
> restart; with(LinearAlgebra):
> w1:=<1,2,3>;   w2:=<1|2|3>;
```

$$w1 := \begin{bmatrix} 1 \\ 2 \\ 3 \end{bmatrix}$$

$$w2 := [1, 2, 3]$$

Note that the column vector is formed by using the comma as a separator while the row vector is formed with the vertical bar separator. We can also declare the vectors using the **Vector** command as shown below.

```
> w1:=Vector([1,2,3]);   w2:=Vector[row]([1,2,3]);
```

$$w1 := \begin{bmatrix} 1 \\ 2 \\ 3 \end{bmatrix}$$

$$w2 := [1, 2, 3]$$

Note that the default for the **Vector** command is a column vector. For matrices we have

```
> A1:=<<1|2>,<3|4>>;   A2:=<<1,2>|<3,4>>;
```

$$A1 := \begin{bmatrix} 1 & 2 \\ 3 & 4 \end{bmatrix}$$

$$A2 := \begin{bmatrix} 1 & 3 \\ 2 & 4 \end{bmatrix}$$

Note that A1 is entered by rows while A2 is entered by columns. We can also use the matrix command where the data is entered by rows.

```
>A1:=Matrix([[1,2,4],[3,4,7]]);
>A2:=Matrix([[1,3],[2,4],[4,7]]);
```

$$A1 := \begin{bmatrix} 1 & 2 & 4 \\ 3 & 4 & 7 \end{bmatrix}$$

$$A2 := \begin{bmatrix} 1 & 3 \\ 2 & 4 \\ 4 & 7 \end{bmatrix}$$

```
> A3:=Matrix(1..3,1..4,-7);
```

$$A3 := \begin{bmatrix} -7 & -7 & -7 & -7 \\ -7 & -7 & -7 & -7 \\ -7 & -7 & -7 & -7 \end{bmatrix}$$

Note that the first entry 1..3 specifies the number of rows in the matrix while the second entry 1..4 specifies the number of columns. The third entry –7 is the value filled into the matrix positions. It has a default value of zero. –7 can be replaced by any other desired value.

We can combine matrices and or matrices and vectors by stacking or appending. In the examples below be sure and note the use of , for stacking and | for side by side assembly.

```
> A4:=<A1,w2>; A5:=<A2|w1>;
```

$$A4 := \begin{bmatrix} 1 & 2 & 4 \\ 3 & 4 & 7 \\ 1 & 2 & 3 \end{bmatrix}$$

$$A5 := \begin{bmatrix} 1 & 3 & 1 \\ 2 & 4 & 2 \\ 4 & 7 & 3 \end{bmatrix}$$

```
> A6:=<A3|A2>;
```

$$A6 := \begin{bmatrix} -7 & -7 & -7 & -7 & 1 & 3 \\ -7 & -7 & -7 & -7 & 2 & 4 \\ -7 & -7 & -7 & -7 & 4 & 7 \end{bmatrix}$$

We shall refer to any number with a decimal point in it as a floating point number. If any entry in the matrix is a floating point number then all the entries are kept as floating point numbers. This is important to remember for accuracy. The most accurate results will be obtained by keeping all of the entries in rational form. Calculations involving matrices with rational entries are done with software routines and the size of the numbers can become quite large. For matrices with floating point entries the calculations are done in the hardware and are limited to approximately 15 digits of accuracy. This last feature can be changed by requiring that the floating point calculations be done with software routines allowing for a larger number of digits. This is done by Setting **UseHardwareFloats:=false;** See the help page **?UseHardwareFloats.** The matrices can have symbolic entries, but since it is designed to be used for computational linear algebra problems we will use real entries. Also we will make frequent use of the command **RandomMatrix.**

The basic operations are those of multiplication of a matrix by a constant, adding two matrices and multiplying matrices together. We begin by defining matrices. The maple commands are shown in red while maple's output is shown in blue.

```
>A:=RandomMatrix(2,2,outputoptions=[datatype=float]);
```

$$A := \begin{bmatrix} 46. & 9. \\ -8. & 93. \end{bmatrix}$$

```
> B:=RandomMatrix(2,2,outputoptions=[datatype=float]);
```

$$B := \begin{bmatrix} -85. & 76. \\ -96. & -51. \end{bmatrix}$$

```
> C:=RandomMatrix(2,2,outputoptions=[datatype=float]);
```

$$C := \begin{bmatrix} -6. & 20. \\ -41. & -37. \end{bmatrix}$$

```
> D1:=RandomMatrix(3,2,outputoptions=[datatype=float]);
```

$$D1 := \begin{bmatrix} -88. & 92. \\ -33. & -98. \\ -62. & -1. \end{bmatrix}$$

```
> E:=RandomMatrix(2,3,outputoptions=[datatype=float]);
```

$$E := \begin{bmatrix} -30. & -72. & -26. \\ -13. & 45. & 70. \end{bmatrix}$$

The name D1 is used in place of D because D is a reserved word in maple. The use of D would generate an error message. Now for some elementary matrix operations.

```
> A1:=5*A; A2:=A+B; A3:=A.E;
```

$$A1 := \begin{bmatrix} 230. & 45. \\ -40. & 465. \end{bmatrix}$$

$$A2 := \begin{bmatrix} -39. & 85. \\ -104. & 42. \end{bmatrix}$$

$$A3 := \begin{bmatrix} -1497. & -2907. & -566. \\ -969. & 4761. & 6718. \end{bmatrix}$$

The only item out of the ordinary in the above computation is the use of the period between A and E in the computation of A3. The period is used for matrix multiplication and also for vector/vector and matrix/vector products. All that is required is that the dimensions and type be compatible. If they are not maple issues an error message.

In using the LinearAlgebra package we need to use a row or column vector in the correct places. If A is a matrix, x a column vector, and y a row vector (of appropriate sizes) then A.x will give a correct result while A.y will generate an error message. Similarly y.A will give correct results while x.A will generate an error message.

```
> A4:=Transpose(E); # A4 is the transpose of E
```

$$\begin{bmatrix} -30. & -13. \\ -72. & 45. \\ -26. & 70. \end{bmatrix}$$

Note that the first row of E is now the first column of A4 while the second row of E is now the second row of A4. For some linear combinations consider:

```
> A5:=5*A-7*B;
```

$$A5 := \begin{bmatrix} 825. & -487. \\ 632. & 822. \end{bmatrix}$$

```
> A6:=17*Transpose(D1)-15*E;
```

$$A6 := \begin{bmatrix} -1046. & 519. & -664. \\ 1759. & -2341. & -1067. \end{bmatrix}$$

```
> A7:=B.(17*Transpose(D1)-15*E);
```

$$A7 := \begin{bmatrix} 222594. & -222031. & -24652. \\ 10707. & 69567. & 118161. \end{bmatrix}$$

We can multiply a vector by a constant and add two vectors together which gives us the ability to form linear combinations. For example

```
> x:=<1|2|3|4>;   y:=<-2|0|4|-7>;
```

$$x := [1, 2, 3, 4]$$

$$y := [-2, 0, 4, -7]$$

```
> w1:=5*x;
```

$$w1 := [5, 10, 15, 20]$$

```
> w2:=x+y;
```

$$w2 := [-1, 2, 7, -3]$$

```
> w3:=4*x-17*y;
```

$$w3 := [38, 8, -56, 135]$$

2. Important Vector Concepts

There are additional vector operations. Let $x = [x_1, x_2, x_3]^T$ where again T means transpose from a row vector to a column vector and vice versa. The vector x has n components with x_1 being the first component and x_n being the nth component. One important property of a vector is its length which is defined to be $\|x\| = \left\{\sum_{k=1}^{n} x_k^2\right\}^{1/2}$. This can be computed using the **Norm** function. If the length is one then the vector is said to be a unit vector. Another important operation is that of the inner product of two vectors usually denoted by $x \circ y$. If $y = [y_1, y_2, \ldots, y_n]^T$ then $x \circ y = \sum_{k=1}^{n} x_k y_k$, For the case of $n \le 3$ we have $x \circ y = \sum_{k=1}^{n} x_k y_k = \|x\|\|y\|\cos\theta$ where θ is the angle between the two vectors x and y. The vectors x and y are said to be orthogonal (that is perpendicular) if $x \circ y = 0$ For the case of $n \le 3$ this means that the angle θ between the vectors x and y is $\frac{\pi}{2}$. Two vectors are parallel that is have the same or exactly opposite directions if there is a constant λ such that x = λy. The following examples illustrate these ideas.

The very first operation is to read in the linear algebra package and note that the command is terminated with a colon to suppress unwanted output.

```
> with(LinearAlgebra):
Warning, the protected names norm and trace have been redefined and
unprotected
```

```
> x:=<1|2|3|4>;  y:=<-2|0|4|-7>;
```

$$x := [1, 2, 3, 4]$$

$$y := [-2, 0, 4, -7]$$

```
> w4:=Norm(x,2);    w5:=Norm(y,2);
```

$$w4 := \sqrt{30}$$

$$w5 := \sqrt{69}$$

```
> w6:=x.y;
```

$$w5 := -18$$

Example 2.1

Determine if the vectors $x = [1,-1,2]$, $y = [1,1,0]$, $u = [4,4,-8]$ are orthogonal, parallel, or neither.

Solution

```
> x:=<1|-1|2>; y:=<1|1|0>; u:=<4|4|-8>;
```

$$x := [1, -1, 2]$$

$$y := [1, 1, 0]$$

$$u := [4, 4, -8]$$

```
> zero:=<0|0|0>;
```

$$zero := [0, 0, 0]$$

```
> a1:=x.y; # x is orthogonal to y x.y = 0
```

$$a1 := 0$$

```
> a2:=x.u; # x is not orthogonal to u
```

$$a2 := -16$$

```
> a3:=y.u; # y is not orthogonal to u
```

$$a3 := 8$$

```
> y1:=evalm(lambda*y);
```

$$y1 := [\lambda, \lambda, 0]$$

We used the command **evalm** to force the multiplication of the vector by the symbol λ. Now we shall test to see if x and y are parallel by seeing if there is a λ that will make $x - y1 = 0$.

```
> w1:=evalm(x-y1)=zero; # Note that w1 is an equation
```

$$w1 := [-\lambda + 1, -\lambda - 1, 2] = [0, 0, 0]$$

They are clearly not parallel for no choice will make the third component, 2, zero. Using the same technique on the other vectors we have

```
> u1:=evalm(lambda*u);
```

$$u1 := [4\,\lambda, 4\,\lambda, -8\,\lambda]$$

```
> w2:=evalm(x-u1)=zero;
```

$$w2 := [-4\,\lambda + 1, -4\,\lambda - 1, 8\,\lambda + 2] = [0, 0, 0]$$

x and y are not parallel for $\lambda = 1/4$ is required to make the first component 0 while $\lambda = -1/4$ is required to make the third component 0. Finally we check y and u.

```
> w3:=evalm(y-u1)=zero;
```

$$w3 := [-4\,\lambda + 1, -4\,\lambda + 1, 8\,\lambda] = [0, 0, 0]$$

$\lambda = 0$ will make the third component 0 but will not make the first or second component 0 hence y and u are not parallel.

Next we consider operations that involve both matrices and vectors. Let

```
> A:=Matrix([[1,2],[3,4],[5,6]]);
```

$$A := \begin{bmatrix} 1 & 2 \\ 3 & 4 \\ 5 & 6 \end{bmatrix}$$

```
> B:=Matrix([[-2,1],[3,-5]]);C:=Matrix([[4,-1],[7,-8]]);
```

$$B := \begin{bmatrix} -2 & 1 \\ 3 & -5 \end{bmatrix}$$

$$C := \begin{bmatrix} 4 & -1 \\ 7 & -8 \end{bmatrix}$$

```
> F:=Matrix([[1,-2,0],[4,0,-7]]);
```

$$F := \begin{bmatrix} 1 & -2 & 0 \\ 4 & 0 & -7 \end{bmatrix}$$

```
> x:=Vector[row]([1,-3,2]); y:=Vector[row]([7,0,-2]);
```

$$x := [1, -3, 2]$$

$$y := [7, 0, -2]$$

```
> u:=Vector[row]([6,3,-1]);
```

$$u := [6, 3, -1]$$

Example 2.2

Determine the value of the linear combination A(3B-5C)F(16x-18y+3u).

Solution

```
> w:=16*x-18*y+3*u;
```

$$w := [-92, -39, 65]$$

```
> G:=A.(3*B-5*C).F;
```

$$G := \begin{bmatrix} 154 & 156 & -406 \\ 314 & 364 & -868 \\ 474 & 572 & -1330 \end{bmatrix}$$

```
> answer:=G.Transpose(w);
```

$$answer := \begin{bmatrix} -46642 \\ -99504 \\ -152366 \end{bmatrix}$$

Example 2.3

For what values of p and q, if any, will A = B where $A = \begin{bmatrix} 2 & 3p-4q \\ 1 & 5 \end{bmatrix}$ and $B = \begin{bmatrix} 2 & 4p+5q \\ q & 5 \end{bmatrix}$?

Solution

```
> A:=Matrix([[2,3*p-4*q],[1,5]]);
```

$$A := \begin{bmatrix} 2 & 3p-4q \\ 1 & 5 \end{bmatrix}$$

```
> B:=Matrix([[2,4*p+5*q],[q,5]]);
```

$$B := \begin{bmatrix} 2 & 4p+5q \\ q & 5 \end{bmatrix}$$

```
> C:=A-B;
```

$$C := \begin{bmatrix} 0 & -p-9q \\ 1-q & 0 \end{bmatrix}$$

Clearly the entry in row 2 and column 1 (position [2,1]) requires q = 1. Armed with this then the entry [1,2] requires -p - 9 = 0 or p = -9. Thus there is only one set of values that will make the two matrices equal.

3. Vector Projection

In physics and engineering forces are frequently used. In many cases the given force must be broken up into the sum of two mutually perpendicular forces whose sum is the given vector. Furthermore one of the two perpendicular forces must have a prescribed direction that is be parallel to a given vector. This problem is most easily solved by looking at the particular problem of finding the projected image of one vector, say a, onto a specified vector say b. This is illustrated in the figure below.

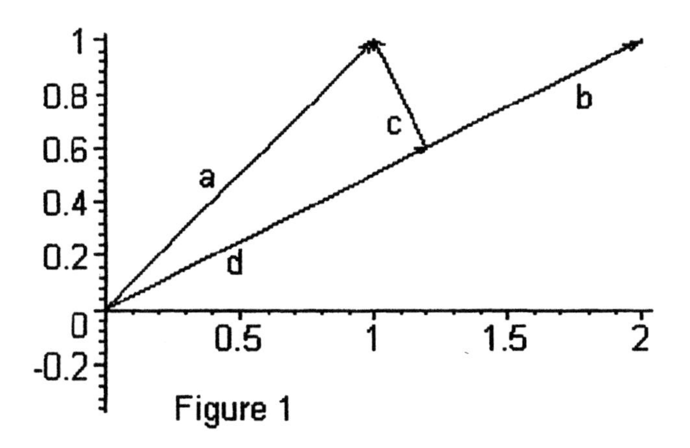

Figure 1

Here a is the given vector which is to be represented as a = d + c where d is parallel to the specified direction b and where c is perpendicular to d. To solve this problem let θ be the angle between the vectors a and b then we have

$$\|d\| = \|a\|\cos(\theta)$$

$$= \|a\| \frac{a \circ b}{\|a\|\|b\|}$$

$$= \frac{a \circ b}{\|b\|}$$

This last result is the length of the vector d. All that we need to do now is to create a unit vector in the direction of b and multiply it by the length of the vector d as in

$d = \frac{a \circ b}{\|b\|} \frac{b}{\|b\|} = \left(\frac{a \circ b}{b \circ b} \right) b$. Once we have d we can find c as c = a – d.

Example 3.1

Given the vectors $x = [4, -5, 7]^T$ and $y = [-8, -17, 2]^T$ represent x and the sum of two mutually perpendicular vectors one of which is parallel to y.

Solution

```
> restart; with(LinearAlgebra):
> x:=Vector[row]([4,-5,7]);     y:=Vector[row]([-8,-17,2]);
```

$$x := [4, -5, 7]$$

$$y := [-8, -17, 2]$$

```
> a1:=x.y;     a2:=y.y;
```

$$a1 := 67$$

$$a2 := 357$$

```
> z1:=(a1/a2)*y;  z2:=x-z1;  # x = z1 + z2
```

$$z1 := \left[\frac{-536}{357}, \frac{-67}{21}, \frac{134}{357}\right]$$

$$z2 := \left[\frac{1964}{357}, \frac{-38}{21}, \frac{2365}{357}\right]$$

Now z1 and z2 are perpendicular by construction. Just as an exercise we shall verify this. We have

```
> w:=z1.z2;
```

$$w := 0$$

Example 3.2

In the figure below the box is being pushed up the ramp with the given force. The ramp is a ten percent grade, that is for every foot forward the ramp rises 1/10 th of a foot. The given force is F = [300,-50]. Break this force up into the sum of two forces F = A + B where A is parallel to the ramp and B is perpendicular to the ramp.

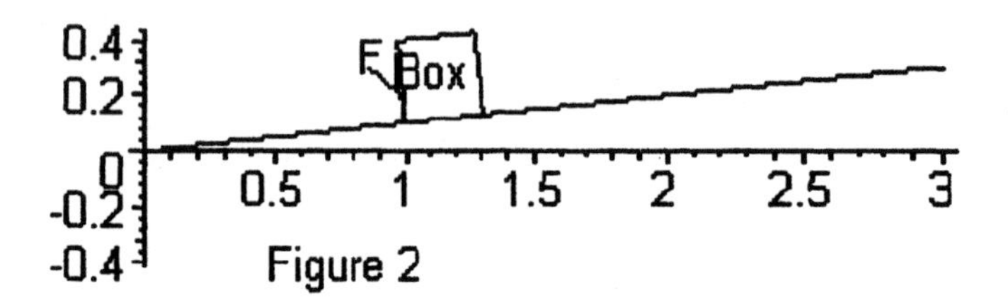

Figure 2

Solution

A vector in the direction of the ramp is R = [1,0.1]. The vector A is then the projection of F onto R. We have

```
> R:=Vector[row]([1,0.1]);  F:=Vector[row]([300,-50]);
```

$$R := [1, 0.1]$$

$$F := [300, -50]$$

```
> a1:=F.R;   a2:=R.R;
```

$$a1 := 295.$$

$$a2 := 1.01000000000000001$$

```
> A:=a1*R/a2;
```

$$A := [292.079207900000028, 29.2079207900000029]$$

```
> B:=F-A;
```
$B := [7.92079209999997147, -79.2079207900000029]$

4. Graphs and Incidence Matrices

For another application of matrices we turn to graph theory. A graph is a set of nodes and edges which connect some of the nodes with a path (here a line segment). Consider the graph below in figure 3.

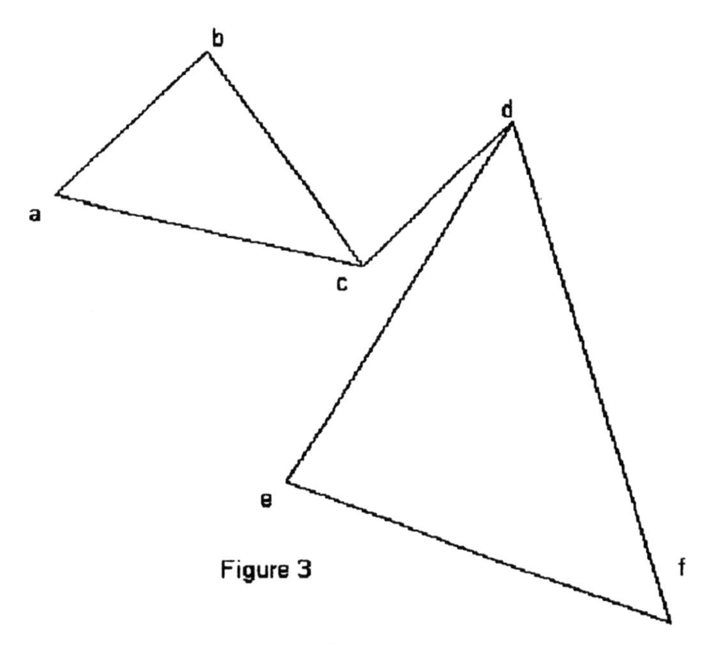

Figure 3

We can represent this graph by means of its incidence matrix A. In the [i,j] entry of the incidence matrix we place a one if there is a path (line segment) joining nodes i and j otherwise we set it equal to 0. The incidence matrix is then

$$A = \begin{array}{c} \begin{matrix} a & b & c & d & e & f \end{matrix} \\ \begin{bmatrix} 0 & 1 & 1 & 0 & 0 & 0 \\ 1 & 0 & 1 & 0 & 0 & 0 \\ 1 & 1 & 0 & 1 & 0 & 0 \\ 0 & 0 & 1 & 0 & 1 & 1 \\ 0 & 0 & 0 & 1 & 0 & 1 \\ 0 & 0 & 0 & 1 & 1 & 0 \end{bmatrix} \end{array} \begin{matrix} a \\ b \\ c \\ d \\ e \\ f \end{matrix}$$

Here we have identified a with 1, b with 2, etc. Thus A[3,4] is the connection, if any, between nodes c and d. Since there is a path between them A[3,4] = 1. As we see the incidence matrix contains exactly the same information as does the original graph. How can we take advantage of this? To make use of the incidence matrix we need to recall how matrix multiplication is done. If

$A = \left[a_{ij}\right]_{m\times n}$ and $B = \left[b_{ij}\right]_{n\times r}$ then $AB = \left[\sum_{k=1}^{n} a_{ik}b_{kj}\right]_{m\times r}$. Now

AA[i,j]= $a_{i1}a_{1j}+a_{i2}a_{2j}+a_{i3}a_{3j}+\cdots+a_{in}a_{jn}$. If $a_{ik}a_{kj}$=1 then there is a path from node i to node k and a path from node k to node j thus there is a path of length 2 from node i to node j. AA[i,j] is the total number of paths of length 2 from node i to node j. Similarly AAA[i,j] is the total number of paths of length 3 from node i to node j. In general $A^k[i,j]$ is the total number of paths of length k from node i to node j. Let us compute some powers of the incidence matrix for the above graph.

```
>
>A:=Matrix([[0,1,1,0,0,0],[1,0,1,0,0,0],[1,1,0,1,0,0],[0,0,1
,0,1,1],[0,0,0,1,0,1],[0,0,0,1,1,0]]);
```

$$A := \begin{bmatrix} 0 & 1 & 1 & 0 & 0 & 0 \\ 1 & 0 & 1 & 0 & 0 & 0 \\ 1 & 1 & 0 & 1 & 0 & 0 \\ 0 & 0 & 1 & 0 & 1 & 1 \\ 0 & 0 & 0 & 1 & 0 & 1 \\ 0 & 0 & 0 & 1 & 1 & 0 \end{bmatrix}$$

```
>A2:=A^2;
```

$$A2 := \begin{bmatrix} 2 & 1 & 1 & 1 & 0 & 0 \\ 1 & 2 & 1 & 1 & 0 & 0 \\ 1 & 1 & 3 & 0 & 1 & 1 \\ 1 & 1 & 0 & 3 & 1 & 1 \\ 0 & 0 & 1 & 1 & 2 & 1 \\ 0 & 0 & 1 & 1 & 1 & 2 \end{bmatrix}$$

AA[3,3] = 3 tells us that there are 3 paths of length 2 joining nodes c to c. They are c-a-c, c-b-c, and c-d-c. AA[3,5] = 1 tells us that there is one path of length 2 joining nodes c and e it is c-d-e.

```
>A4:=A^4;
```

$$A4 := \begin{bmatrix} 7 & 6 & 6 & 6 & 2 & 2 \\ 6 & 7 & 6 & 6 & 2 & 2 \\ 6 & 6 & 13 & 4 & 6 & 6 \\ 6 & 6 & 4 & 13 & 6 & 6 \\ 2 & 2 & 6 & 6 & 7 & 6 \\ 2 & 2 & 6 & 6 & 6 & 7 \end{bmatrix}$$

AAAA[6,1] = 2 tells us that there are 2 paths of length 4 joining nodes f and a, they are a-c-d-e-f and a-b-c-d-f.

5. Markov Chains

In our next example we shall look at an example of a Markov chain. A Markov matrix is one in which every entry satisfies $0 \le a_{ij} \le 1$ and where the sum of the entries in each column is one. For this example we shall set C = cloudy, R = rainy, and S = sunny. We assume that C, R, and S are the only weather possibilities. Set

$$\begin{array}{c} \text{today} \\ \begin{array}{ccc} S & C & R \end{array} \end{array}$$

$$A=\begin{bmatrix} \frac{5}{8} & \frac{3}{8} & \frac{5}{8} \\ \frac{1}{4} & \frac{3}{8} & \frac{1}{4} \\ \frac{1}{8} & \frac{1}{4} & \frac{1}{8} \end{bmatrix} \begin{array}{l} S \\ C \\ R \end{array} \text{ tomorrow}$$

We interpret the matrix A as follows. If it is sunny today then the probability that it will be cloudy tomorrow is A[2,1] = 1/4 or 0.25. If it is cloudy today then the probability that it will be sunny tomorrow is A[1,2] = 3/8.We enter the data into the matrix in rational form because this will give more accurate results when we compute powers of A. If we had entered the results in decimal form we would have gotten less accurate results when computing powers of A.

Now let S0 = [1/2,1/4,1/4] be the probabilities of S, C, and R for today's weather. The probability for tomorrows weather then is S1 = A(S0). The probability for the weather the day after tomorrow will be S2 = A(S1) = AA(S0) etc. Now for some computations.

```
> A:=Matrix([[5/8,3/8,5/8],[1/4,3/8,1/4],[1/8,1/4,1/8]]);
```

$$A := \begin{bmatrix} \frac{5}{8} & \frac{3}{8} & \frac{5}{8} \\ \frac{1}{4} & \frac{3}{8} & \frac{1}{4} \\ \frac{1}{8} & \frac{1}{4} & \frac{1}{8} \end{bmatrix}$$

```
> S0:=<1/2,1/4,1/4>;
```

$$S0 := \begin{bmatrix} \frac{1}{2} \\ \frac{1}{4} \\ \frac{1}{4} \end{bmatrix}$$

```
> S1:=A.S0;
```

$$S1 := \begin{bmatrix} \frac{9}{16} \\ \frac{9}{32} \\ \frac{5}{32} \end{bmatrix}$$

```
> S2:=A.S1;
```

$$S2 := \begin{bmatrix} \frac{71}{128} \\ \frac{73}{256} \\ \frac{41}{256} \end{bmatrix}$$

Now lets see if there is a long term trend. Lets compute the 100th and 101st power of A and apply it to S0 to see if the sequence of S's is converging to anything. We are also going to want to see the resulting entries in the vector in decimal form. To convert we will use the **map** function. The symbol \ is used to denote the continuation of a number in maple.

```
> S100:=(A^100).S0; S100A:=map(evalf,S100);
```

S100 :=

[2255289830941752452654350583596097249735554292987286563204298354 \
6423507247379799247030470954 / 40740719526689721725368913768187564 \
32210293678733187250127228089870876259952667341236679475 2]

[23280411158108412414496522153250036126302495927610700007270176564 \
40500719972952766420959700 1 / 81481439053379443450737827536375124 \
64420587357466374500254456179741752519905334682473358950 4]

[13095231276435981983154293711203145321045153959281018754089474314 \
7278165498478593111178983134 / 81481439053379443450737827536375124 \
64420587357466374500254456179741752519905334682473358950 4]

$$S100A := \begin{bmatrix} 0.5535714286 \\ 0.2857142857 \\ 0.1607142857 \end{bmatrix}$$

Note that the entries in the vector S100 are in rational form and hence are more accurate than their ten digit approximations in the vector S100A.

```
> S101:=evalm(A&*S100); S101A:=map(evalf,S101);
```

S101 := [18042318647534019621234804668768777997884434343898292505634 38 \
6837138805797903839397624376759 / 3259257562135177738029513101 45 \
50050576823494298654980010178247189670100796213387298934358016 , 186\
2432892648672993159721772260002890104199674208856000581614125124 0 \
0575978362213136767760 09 / 651851512427035547605902620291001011 5 \
3646988597309960020356494379340201592426774597868716032 , 1047618502\
1148785586523434968962516256836123167424815003271579453822532398 7 \
82874488943186505 / 651851512427035547605902620291001011536469 88 \
5973099600203564943793402015924267745978687160 32]

S101A := [0.5535714286, 0.2857142857, 0.1607142857]

To see how close together these last two vectors are, we shall look at the difference w:=S101 - S100

```
>w:=evalm(S101-S100); w1:=map(evalf,w);
>w2:=evalm(S101A-S100A);
```

w := [-1/32592575621351777380295131014550050576823494298654980010178 24 \
718967010079621338729893435801 6, 1/651851512427035547605902620291 0\
0101153646988597309960020356494379340201592426774597868716032 , 1/6\
5185151242703554760590262029100101153646988597309960020356494379 3 \
402015924267745978687160 32]

$w1 := [-0.3068183416\ 10^{-91}, 0.1534091708\ 10^{-91}, 0.1534091708\ 10^{-91}]$

$w2 := [0., 0., 0.]$

This result shows that the vectors are nearly equal and so the sequence of S's seems to be converging. Note that the difference in the vectors S100 and S101 does not result in the zero vector while subtracting S101A from S100A does result in the zero vector. This shows that the rational entries are more accurate than the ten digit floating point entries.

6. Complex Values

Maple can handle complex as well as real values in both vectors and matrices. We enter complex values as 3 + 4*I where I stands for the square root of -1. For some examples we have:

```
> a:=2+3*I;   b:=4-5*I;
```

$$a := 2 + 3\,I$$

$$b := 4 - 5\,I$$

```
> w:=a*b;
```

$$w := 23 + 2\,I$$

```
> a1:=conjugate(a); w1:=a1*b;
```

$$a1 := 2 - 3\,I$$

$$w1 := -7 - 22\,I$$

For complex valued vectors the inner product of two vectors x and y is defined to be $x \circ y = \sum_{k=1}^{n} \bar{x}_k y_k$ where the horizontal bar over x_k indicates the conjugate of x_k. This definition reduces to the one given earlier for real vectors. The maple function **DotProduct** applies this definition and so the inner product of complex vectors will be computed correctly. For some examples we have:

```
> A:=Matrix([[2+3*I,4-5*I],[6+7*I,1-2*I]]);
```

$$A := \begin{bmatrix} 2+3\,I & 4-5\,I \\ 6+7\,I & 1-2\,I \end{bmatrix}$$

```
> x:=Vector([1-7*I,13+21*I]); y:=Vector([2-5*I,17+15*I]);
```

$$x := \begin{bmatrix} 1-7\,I \\ 13+21\,I \end{bmatrix}$$

$$y := \begin{bmatrix} 2-5\,I \\ 17+15\,I \end{bmatrix}$$

```
> B:=A.x;
```

$$B := \begin{bmatrix} 180+8\,I \\ 110-40\,I \end{bmatrix}$$

```
> w:=x.y;
```

$$w := 573 - 153\,I$$

Exercises

Given the matrices;

$$A = \begin{bmatrix} -4 & 2 \\ 5 & -6 \end{bmatrix}, \quad B = \begin{bmatrix} 7 & -3 \\ 2 & 10 \end{bmatrix}, \quad C = \begin{bmatrix} 15 & 12 \\ -18 & 26 \end{bmatrix},$$

$$F=\begin{bmatrix}3 & 0 & 17 & 21\\ 0 & -7 & 9 & -23\end{bmatrix},\quad G=\begin{bmatrix}1 & 3\\ 2 & 7\\ -5 & 12\\ 0 & 18\end{bmatrix}$$ and the vectors

$x=[2,-3]^T,\ \ y=[-5,8]^T,\ \ z=[12,21]^T$

$u=[2,5,-8,17]^T,\ \ v=[-18,21,0,-6]^T$. Solve the following problems:

1). Find Fv.
2). Find Ax.
3). Find the linear combination 3x - 17y + 21z.
4). Find the linear combination 5A – 8B + 25C.
5). Find the matrix products AG^T.
6). Find the matrix product FG.
7). Find the matrix/vector product Gx.
8). Find the matrix /vector product Fu.
9). Find the linear combination G(Fv + C(5A – 127B)(17x – 32y + 25z)).
10). Write the vector u as the sum of two mutually perpendicular vectors one of which is parallel to v.
11). Write the vector v as the sum of two mutually perpendicular vectors one of which is parallel to u.
12). For the graph shown above determine the total number of paths of length ≤ 6 between any two nodes.
13). For the graph shown above determine how many paths of length 10 exist between nodes a and f.
14). For the graph shown above determine how many paths of length 15 exist between nodes b and e.
15). For the Markov weather matrix determine the terms in the sequence S1, S2, …, S10 starting with the given S0
16). Starting with S0 generate the terms in the sequence S1, S2, …, Sn, … Ineach case compute the differences S2-S1, S3-S2, etc until all the entries in the difference vector are $\leq 10^{-4}$. What is the smallest value of n that will work?

Chapter 3.0

Solving Systems of Equations

1. Equation Reduction and the Augmented Matrix

In chapter 2 we saw that the system of equations

$$a_{11}x_1 + a_{12}x_2 + a_{13}x_3 = b_1$$
$$a_{21}x_1 + a_{22}x_2 + a_{23}x_3 = b_2$$
$$a_{31}x_1 + a_{32}x_2 + a_{33}x_3 = b_3$$

could be written as the matrix equation Ax = b where $A = \begin{bmatrix} a_{11} & a_{12} & a_{13} \\ a_{21} & a_{22} & a_{23} \\ a_{31} & a_{32} & a_{33} \end{bmatrix}$,

$b = [b_1, b_2, b_3]^T$ and $x = [x_1, x_2, x_3]^T$. There are three operations that we can perform on the system of equations without changing the solution. They are:

1). Interchange two equations.

2). Multiply an equation by a non zero scalar (constant).

3). Multiply equation j by a non zero scalar and add the result to equation m where j does not equal m.

Now consider the augmented matrix $[A \mid b] = \begin{bmatrix} a_{11} & a_{12} & a_{13} & b_1 \\ a_{21} & a_{22} & a_{23} & b_2 \\ a_{31} & a_{32} & a_{33} & b_3 \end{bmatrix}$. This matrix contains all the essential information in the system of equations. The three operations given above are now performed on the rows of [A|b]. The object is to reduce this augmented matrix to either row echelon form or reduced row echelon form. Once we have one of these two forms we can easily solve the system of equations.

In our reductions below we shall use quasi echelon forms in the sense that we will not insist on having a one in the usual position but will be happy with any non zero value.

2. Reduction to Echelon Form

The maple commands that will be useful for this section are **addrow, mulrow, pivot, gausselim, gaussjord,** and **linslove.** Maple has a single command that will solve the system of equations Ax = b. The maple command is **linsolve.** Nearly all Linear Algebra texts begin by solving these equations using the three rules stated at the beginning of section 1. For this reason then we shall first solve the system of equations by the rather long and tedious method that it is shown in your text. We shall use the commands **randmatrix** and **randvector** to generate random matrices and random vectors.

```
> restart; with(linalg):
```

```
Warning, the protected names norm and trace have been redefined and
unprotected
```

```
> A:=randmatrix(4,4); b:=randvector(4);
```

$$A := \begin{bmatrix} -85 & -55 & -37 & -35 \\ 97 & 50 & 79 & 56 \\ 49 & 63 & 57 & -59 \\ 45 & -8 & -93 & 92 \end{bmatrix}$$

$$b := [43, -62, 77, 66]$$

```
> A1:=augment(A,b);
```

$$A1 := \begin{bmatrix} -85 & -55 & -37 & -35 & 43 \\ 97 & 50 & 79 & 56 & -62 \\ 49 & 63 & 57 & -59 & 77 \\ 45 & -8 & -93 & 92 & 66 \end{bmatrix}$$

```
> A2:=addrow(A1,1,2,97/85);
```

$$A2 := \begin{bmatrix} -85 & -55 & -37 & -35 & 43 \\ 0 & \frac{-217}{17} & \frac{3126}{85} & \frac{273}{17} & \frac{-1099}{85} \\ 49 & 63 & 57 & -59 & 77 \\ 45 & -8 & -93 & 92 & 66 \end{bmatrix}$$

```
> A3:=addrow(A2,1,3,49/85);
```

$$A3 := \begin{bmatrix} -85 & -55 & -37 & -35 & 43 \\ 0 & \frac{-217}{17} & \frac{3126}{85} & \frac{273}{17} & \frac{-1099}{85} \\ 0 & \frac{532}{17} & \frac{3032}{85} & \frac{-1346}{17} & \frac{8652}{85} \\ 45 & -8 & -93 & 92 & 66 \end{bmatrix}$$

At this point I think that we can see that we can reduce the augmented matrix to upper triangular form by continuing to put zeroes into the appropriate spaces. We shall drop the use of **addrow** now and look instead at the command **pivot.** This command is preferable since it will process an entire column rather than a single entry. We have

```
> B1:=pivot(A1,1,1);
```

$$B1 := \begin{bmatrix} -85 & -55 & -37 & -35 & 43 \\ 0 & \frac{-217}{17} & \frac{3126}{85} & \frac{273}{17} & \frac{-1099}{85} \\ 0 & \frac{532}{17} & \frac{3032}{85} & \frac{-1346}{17} & \frac{8652}{85} \\ 0 & \frac{-631}{17} & \frac{-1914}{17} & \frac{1249}{17} & \frac{1509}{17} \end{bmatrix}$$

```
> B2:=pivot(B1,2,2,2..4);
```

$$B2 := \begin{bmatrix} -85 & -55 & -37 & -35 & 43 \\ 0 & \frac{-217}{17} & \frac{3126}{85} & \frac{273}{17} & \frac{-1099}{85} \\ 0 & 0 & \frac{19504}{155} & \frac{-1234}{31} & \frac{10864}{155} \\ 0 & 0 & \frac{-238188}{1085} & \frac{830}{31} & \frac{19586}{155} \end{bmatrix}$$

```
> B3:=pivot(B2,3,3,3..4);
```

$$B3 := \begin{bmatrix} -85 & -55 & -37 & -35 & 43 \\ 0 & \frac{-217}{17} & \frac{3126}{85} & \frac{273}{17} & \frac{-1099}{85} \\ 0 & 0 & \frac{19504}{155} & \frac{-1234}{31} & \frac{10864}{155} \\ 0 & 0 & 0 & \frac{-31663}{742} & \frac{13178}{53} \end{bmatrix}$$

Thus the **pivot** command has rather quickly reduced the augmented matrix to upper triangular (quasi row echelon) form. It can now be easily solved with the maple command **backsub.** This gives

```
> x:=backsub(B3);
```

$$x := \left[\frac{13019547}{1456498}, \frac{-7299476}{728249}, \frac{-1873417}{1456498}, \frac{-184492}{31663} \right].$$

The last parameter in the **pivot** command controls where the zeroes will go. If we use an alternate form of the command we will get the quasi reduced row echelon form as shown in

```
> C1:=pivot(A1,1,1);
```

$$C1 := \begin{bmatrix} -85 & -55 & -37 & -35 & 43 \\ 0 & \frac{-217}{17} & \frac{3126}{85} & \frac{273}{17} & \frac{-1099}{85} \\ 0 & \frac{532}{17} & \frac{3032}{85} & \frac{-1346}{17} & \frac{8652}{85} \\ 0 & \frac{-631}{17} & \frac{-1914}{17} & \frac{1249}{17} & \frac{1509}{17} \end{bmatrix}$$

```
> C2:=pivot(C1,2,2);
```

$$C2 := \begin{bmatrix} -85 & 0 & \frac{-42415}{217} & \frac{-3230}{31} & \frac{3060}{31} \\ 0 & \frac{-217}{17} & \frac{3126}{85} & \frac{-273}{17} & \frac{-1099}{85} \\ 0 & 0 & \frac{19504}{155} & \frac{-1234}{31} & \frac{10864}{155} \\ 0 & 0 & \frac{-238188}{1085} & \frac{830}{31} & \frac{19586}{155} \end{bmatrix}$$

```
> C3:=pivot(C2,3,3);
```

$$C3 := \begin{bmatrix} -85 & 0 & 0 & \frac{-11333645}{68264} & \frac{253045}{1219} \\ 0 & \frac{-217}{17} & 0 & \frac{2295519}{82892} & \frac{-692447}{20723} \\ 0 & 0 & \frac{19504}{155} & \frac{-1234}{31} & \frac{10864}{155} \\ 0 & 0 & 0 & \frac{-31663}{742} & \frac{13178}{53} \end{bmatrix}$$

```
> C4:=pivot(C3,4,4);
```

$$C4 := \begin{bmatrix} -85 & 0 & 0 & 0 & \frac{-1106661495}{1456498} \\ 0 & \frac{-217}{17} & 0 & 0 & \frac{1583986292}{12380233} \\ 0 & 0 & \frac{19504}{155} & 0 & \frac{-794328808}{4907765} \\ 0 & 0 & 0 & \frac{-31663}{742} & \frac{13178}{53} \end{bmatrix}$$

In this case we can easily calculate the answer by using **mulrow** to put ones in the nonzero positions of columns 1, 2, 3, and 4. We have

```
> C5:=mulrow(C4,1,-1/85);
```

$$C5 := \begin{bmatrix} 1 & 0 & 0 & 0 & \frac{13019547}{1456498} \\ 0 & \frac{-217}{17} & 0 & 0 & \frac{1583986292}{12380233} \\ 0 & 0 & \frac{19504}{155} & 0 & \frac{-794328808}{4907765} \\ 0 & 0 & 0 & \frac{-31663}{742} & \frac{13178}{53} \end{bmatrix}$$

```
> C6:=mulrow(C5,2,-17/217);
```

$$C6 := \begin{bmatrix} 1 & 0 & 0 & 0 & \frac{13019547}{1456498} \\ 0 & 1 & 0 & 0 & \frac{-7299476}{728249} \\ 0 & 0 & \frac{19504}{155} & 0 & \frac{-794328808}{4907765} \\ 0 & 0 & 0 & \frac{-31663}{742} & \frac{13178}{53} \end{bmatrix}$$

```
> C7:=mulrow(C6,3,155/19504);
```

$$C7 := \begin{bmatrix} 1 & 0 & 0 & 0 & \frac{13019547}{1456498} \\ 0 & 1 & 0 & 0 & \frac{-7299476}{728249} \\ 0 & 0 & 1 & 0 & \frac{-1873417}{1456498} \\ 0 & 0 & 0 & \frac{-31663}{742} & \frac{13178}{53} \end{bmatrix}$$

```
> C8:=mulrow(C7,4,-742/31663);
```

$$C8 := \begin{bmatrix} 1 & 0 & 0 & 0 & \frac{13019547}{1456498} \\ 0 & 1 & 0 & 0 & \frac{-7299476}{728249} \\ 0 & 0 & 1 & 0 & \frac{-1873417}{1456498} \\ 0 & 0 & 0 & 1 & \frac{-184492}{31663} \end{bmatrix}$$

Finally we look at the maple commands **gausselim** and **gaussjord.** These commands are the fastest way of arriving at the above quasi echelon forms.

```
> D1:=gausselim(A1);
```

$$D1 := \begin{bmatrix} -85 & -55 & -37 & -35 & 43 \\ 0 & \frac{-217}{17} & \frac{3126}{85} & \frac{273}{17} & \frac{-1099}{85} \\ 0 & 0 & \frac{19504}{155} & \frac{-1234}{31} & \frac{10864}{155} \\ 0 & 0 & 0 & \frac{-31663}{742} & \frac{13178}{53} \end{bmatrix}$$

```
> D2:=gaussjord(A1);
```

$$D2 := \begin{bmatrix} 1 & 0 & 0 & 0 & \frac{13019547}{1456498} \\ 0 & 1 & 0 & 0 & \frac{-7299476}{728249} \\ 0 & 0 & 1 & 0 & \frac{-1873417}{1456498} \\ 0 & 0 & 0 & 1 & \frac{-184492}{31663} \end{bmatrix}$$

We can obtain the solution from D1 by applying the **backsub** command. When a system of equations has a unique solution such as in the above case we say that A is nonsingular. When there are infinitely many solutions we say that A is singular. When there are no solutions we say that the system of equations is inconsistent.

3. Singular systems

We can determine that a system of equations is singular from its echelon form. If it is singular then its echelon form will contain one or more rows of zeroes. Consider the following example of an inconsistent system of equations:

```
> A:=matrix(4,4,(i,j)->i+j); b1:=vector([3,-2,0,1]);
```

$$A := \begin{bmatrix} 2 & 3 & 4 & 5 \\ 3 & 4 & 5 & 6 \\ 4 & 5 & 6 & 7 \\ 5 & 6 & 7 & 8 \end{bmatrix}$$

$$b1 := [3, -2, 0, 1]$$

```
> A1:=augment(A,b1);
```

$$A1 := \begin{bmatrix} 2 & 3 & 4 & 5 & 3 \\ 3 & 4 & 5 & 6 & -2 \\ 4 & 5 & 6 & 7 & 0 \\ 5 & 6 & 7 & 8 & 1 \end{bmatrix}$$

```
> A2:=gaussjord(A1);
```

$$A2 := \begin{bmatrix} 1 & 0 & -1 & -2 & 0 \\ 0 & 1 & 2 & 3 & 0 \\ 0 & 0 & 0 & 0 & 1 \\ 0 & 0 & 0 & 0 & 0 \end{bmatrix}$$

It is easy to see that this system of equations is inconsistent because of row 3. The equation corresponding to row 3 is $0x_1 + 0x_2 + 0x_3 + 0x_4 = 1$ for which there is obviously no solution. We can modify the vector b to get an infinite number of solutions as in:

```
> b2:=vector([1,1,1,1]);
```

$$b2 := [1, 1, 1, 1]$$

```
> A3:=augment(A,b2);
```

$$A3 := \begin{bmatrix} 2 & 3 & 4 & 5 & 1 \\ 3 & 4 & 5 & 6 & 1 \\ 4 & 5 & 6 & 7 & 1 \\ 5 & 6 & 7 & 8 & 1 \end{bmatrix}$$

```
> A4:=gaussjord(A3);
```

$$A4 := \begin{bmatrix} 1 & 0 & -1 & -2 & -1 \\ 0 & 1 & 2 & 3 & 1 \\ 0 & 0 & 0 & 0 & 0 \\ 0 & 0 & 0 & 0 & 0 \end{bmatrix}$$

```
> x:=backsub(A4);
```

$$x := [-1 + \_t_2 + 2\,\_t_1,\ 1 - 2\,\_t_2 - 3\,\_t_1,\ \_t_2,\ \_t_1]$$

The occurrence of the parameters $\_t_1$, $\_t_2$ indicate that there are an infinite number of solutions. We can get particular solutions by assigning particular values to the parameters for instance

```
> x1:=map2(subs,{_t[1]=0,_t[2]=0},x);
```

$$x1 := [-1, 1, 0, 0]$$

```
> x2:=map2(subs,{_t[1]=1,_t[2]=0},x);
```

$$x2 := [1, -2, 0, 1]$$

```
>x3:=map2(subs,{_t[1]=1,_t[2]=1},x);
```

$$x3 := [2, -4, 1, 1]$$

4. The inverse Matrix and the LU decomposition

B is the matrix inverse to the matrix A if AB = I and BA = I where I is the identity matrix. The inverse matrix exists if A is nonsingular that is if the solution of the equation Ax = b is unique. When it exists we denote the inverse matrix by the symbol A^{-1}.

Suppose that we have a series of matrix equations to solve such as $Ax_1 = b_1, Ax_2 = b_2, \cdots, Ax_m = b_m$ we can solve all of these simultaneously as follows. LetX be the matrix with columns $x_1, x_2, \ldots, x_m$ and B be the matrix with columns $b_1, b_2, \ldots, b_m$ then the individual problems are contained in the single matrix equation AX = B. This can be solved by reducing the augmented matrix [A|B] to reduced row echelon form. We can apply this observation to find the inverse matrix when it exists. We simply solve the matrix equation AX = I. We can start with the augmented matrix [A|I] and proceed to reduce this to reduced row echelon form which will be $\left[I \mid A^{-1}\right]$. For example

```
> A:=randmatrix(3,3); I1:=diag(1,1,1);
```

$$A := \begin{bmatrix} 77 & 66 & 54 \\ -5 & 99 & -61 \\ -50 & -12 & -18 \end{bmatrix}$$

$$I1 := \begin{bmatrix} 1 & 0 & 0 \\ 0 & 1 & 0 \\ 0 & 0 & 1 \end{bmatrix}$$

```
> B:=augment(A,I1);
```

$$B := \begin{bmatrix} 77 & 66 & 54 & 1 & 0 & 0 \\ -5 & 99 & -61 & 0 & 1 & 0 \\ -50 & -12 & -18 & 0 & 0 & 1 \end{bmatrix}$$

```
> B1:=gaussjord(B);
```

$$B1 := \begin{bmatrix} 1 & 0 & 0 & \frac{-419}{45387} & \frac{10}{5043} & \frac{-1562}{45387} \\ 0 & 1 & 0 & \frac{1480}{136161} & \frac{73}{15129} & \frac{4427}{272322} \\ 0 & 0 & 1 & \frac{835}{45387} & \frac{-44}{5043} & \frac{2651}{90774} \end{bmatrix}$$

To pick out the inverse we can use maple's **submatrix** command. As parameters we provide the underlying matrix name, a list of the rows to keep and finally a list of the columns to keep. This will define the sub matrix that we want to extract.

```
> Ainv:=submatrix(B1,[1,2,3],[4,5,6]);
```

$$Ainv := \begin{bmatrix} \frac{-419}{45387} & \frac{10}{5043} & \frac{-1562}{45387} \\ \frac{1480}{136161} & \frac{73}{15129} & \frac{4427}{272322} \\ \frac{835}{45387} & \frac{-44}{5043} & \frac{2651}{90774} \end{bmatrix}$$

In practice, of course, we use the maple function **inverse** to get the inverse matrix as in

```
> B3:=inverse(A);
```

$$B3 := \begin{bmatrix} \frac{-419}{45387} & \frac{10}{5043} & \frac{-1562}{45387} \\ \frac{1480}{136161} & \frac{73}{15129} & \frac{4427}{272322} \\ \frac{835}{45387} & \frac{-44}{5043} & \frac{2651}{90774} \end{bmatrix}$$

The inverse matrix can be used to solve the matrix equation Ax = b as follows

$$Ax = b \Rightarrow A^{-1}Ax = A^{-1}b$$

$$Ix = A^{-1}b.$$

$$x = A^{-1}b$$

Although this is theoretically acceptable it is not used in solving large systems of equations. Rather various transformations are used instead. One of these is the LU decomposition.

P is a permutation matrix if it is derived from the identity matrix by permuting the rows of the identity matrix. A permutation matrix satisfies the relationship PP = I. The maple command **LUdecomp** will give the following decomposition of a matrix A, A = PLU. Here P is a permutation matrix, L is a unit lower triangular matrix, that is, it has ones along the main diagonal, and U is an upper triangular matrix. Once we have the decomposition we solve Ax = b as PLUx = b and since PP = I this gives LUx = Pb. We then solve this problem as

$$\begin{aligned} Ly &= Pb \\ Ux &= y \end{aligned}$$

This may seem to be a very cumbersome way to solve the problem but actually there are very good numerical reasons for using methods of this type.

Example 4.1

Using the LU decomposition solve the matrix equation Ax = b where

$$A = \begin{bmatrix} 1 & -5 & 3 \\ 2 & 0 & -7 \\ 15 & -3 & 4 \end{bmatrix} \text{ and } b^T = [1,2,3].$$

Solution

```
>A:=matrix([[1,-5,3],[2,0,-7],[15,-3,4]]);
```

$$A := \begin{bmatrix} 1 & -5 & 3 \\ 2 & 0 & -7 \\ 15 & -3 & 4 \end{bmatrix}$$

```
>b:=vector([1,2,3]);
```

$$b := [1, 2, 3]$$

```
>LUdecomp(A,P='P',L='L',U='U');
```

$$\begin{bmatrix} 1 & -5 & 3 \\ 0 & 10 & -13 \\ 0 & 0 & \frac{263}{5} \end{bmatrix}$$

```
>print(P,L,U); # Note that P = I for this problem
```

$$\begin{bmatrix} 1 & 0 & 0 \\ 0 & 1 & 0 \\ 0 & 0 & 1 \end{bmatrix}, \begin{bmatrix} 1 & 0 & 0 \\ 2 & 1 & 0 \\ 15 & \frac{36}{5} & 1 \end{bmatrix}, \begin{bmatrix} 1 & -5 & 3 \\ 0 & 10 & -13 \\ 0 & 0 & \frac{263}{5} \end{bmatrix}$$

```
>y:=forwardsub(augment(L,b));
```

$$y := [1, 0, -12]$$

```
>x:=backsub(augment(U,y));
```

$$x := \left[\frac{53}{263}, \frac{-78}{263}, \frac{-60}{263}\right]$$

Just as a check on our work we have

```
> check:=evalm(A&*x-b);
```

$$check := [0, 0, 0]$$

The zero result shows that we have correctly found the solution to the original problem. For an example where the matrix P is not the identity matrix consider

Example 4.2

Using the LU decomposition solve the matrix equation Ax = b where $A = \begin{bmatrix} 1 & -1 & 3 \\ 2 & -2 & 0 \\ 5 & -3 & 1 \end{bmatrix}$ and $b^T = [1,2,3]$.

Solution

```
> A:=matrix([[1,-1,3],[2,-2,0],[5,-3,1]]);
```

$$A := \begin{bmatrix} 1 & -1 & 3 \\ 2 & -2 & 0 \\ 5 & -3 & 1 \end{bmatrix}$$

```
> b:=vector([1,2,3]);
```

$$b := [1, 2, 3]$$

```
> LUdecomp(A,P='P',L='L',U='U'):
> print(P,L,U);
```

$$\begin{bmatrix} 1 & 0 & 0 \\ 0 & 0 & 1 \\ 0 & 1 & 0 \end{bmatrix}, \begin{bmatrix} 1 & 0 & 0 \\ 5 & 1 & 0 \\ 2 & 0 & 1 \end{bmatrix}, \begin{bmatrix} 1 & -1 & 3 \\ 0 & 2 & -14 \\ 0 & 0 & -6 \end{bmatrix}$$

```
> y:=forwardsub(augment(L,evalm(P&*b)));
```

$$y := [1, -2, 0]$$

```
> x:=backsub(U,y);
```

$$x := [0, -1, 0]$$

```
> check:=evalm(A&*x-b);
```

$$check := [0, 0, 0]$$

5. Linear Independence, Rank, and Determinant

A set of vectors $S = \{v_1, v_2, \ldots, v_m\}$ is said to be linearly independent if $\sum_{k=1}^{m} c_k v_k = 0$ can hold only if each of the constants $c_k = 0$. Otherwise they are linearly dependent. This idea is not unique to vectors, for instance we can let the v's be matrices of the same size. Linear independence is an idea that applies to a wide variety of situations, In Linear

Algebra we are primarily concerned with vectors and shall stick with them. Let A be the matrix whose first column is v_1, whose second column is v_2 etc. Then the vectors in the set S are linearly independent if and only if the equation Ac = 0 has the unique solution c = 0. For sets of vectors this is an easy test to apply.

Example 5.1

Determine if the set S of vectors given below is linearly independent or not.

$S = \{v_1 = [1,2,-1,3], v_2 = [2,0,3,-1], v_3 = [3,2,2,2]\}$.

Solution

```
> A:=matrix([[1,2,3],[2,0,2],[-1,3,2],[3,-1,2]]);
```

$$A := \begin{bmatrix} 1 & 2 & 3 \\ 2 & 0 & 2 \\ -1 & 3 & 2 \\ 3 & -1 & 2 \end{bmatrix}$$

```
> b:=vector([0,0,0,0]);
```

$$b := [0, 0, 0, 0]$$

```
> c:=linsolve(A,b);
```

$$c := [-\_t_1, -\_t_1, \_t_1]$$

From the answer for c it is clear that there are an infinite number of solutions to the equation, in particular [-1,-1,1], and hence the vectors are linearly dependent. Another important property of a matrix is its rank. The rank of a matrix A is the number of linearly independent columns in A. The number of linearly independent columns is always the same as the number of linearly independent rows in A.

In the above example we could have answered the question by finding the rank of A, it is 2. The fact that the rank is less than the number of columns tells us that the columns are linearly dependent. In an m by n matrix A the columns are linearly independent if and only if the rank is n. The maple command to determine the rank is **rank.**

Example 5.2

In this example we shall revisit an example of a Markov chain from chapter 2. A Markov matrix is one in which every entry satisfies $0 \le a_{ij} \le 1$ and where the sum of the entries in each column is one. For this example we shall set C = cloudy, R = rainy, and S = sunny. We assume that C, R, and S are the only weather possibilities. Set

$$A=\begin{bmatrix} \frac{5}{8} & \frac{3}{8} & \frac{5}{8} \\ \frac{1}{4} & \frac{3}{8} & \frac{1}{4} \\ \frac{1}{8} & \frac{1}{4} & \frac{1}{8} \end{bmatrix}$$

(columns: today S, C, R; rows: tomorrow S, C, R)

We interpret the matrix A as follows. If it is sunny today then the probability that it will be cloudy tomorrow is A[2,1] = 1/4 or 0.25. If it is cloudy today then the probability that it will be sunny tomorrow is A[1,2] = 3/8.We enter the data into the matrix in rational form because this will give more accurate results when we compute powers of A.

Now let S0 = [1/2,1/4,1/4] be the probabilities of S, C, and R for today's weather. The probability for tomorrows weather then is S1 = A(S0). The probability for the weather the day after tomorrow will be S2 = A(S1) = AA(S0) etc. The question that we want to answer is does the sequence of vectors S0, S1, S2,... approach a limiting value? In the example in chapter 2 we saw that the answer was probably yes. How can we find the limiting vector? If we denote the limiting vector by S then we have AS = S. This is the equation whose solution, if it exists, is the limiting vector. Solutions of this type are called fixed points. We can solve the matrix equation by noting that AS – IS = 0 where I is the identity matrix. Thus we must solve (A-I)S = 0. Before solving the equation we need to note that we are looking for a nontrivial solution S. Furthermore the entries in S must all add up to 1. With that in mind then we can now solve the equation

```
> A:=matrix([[5/8,3/8,5/8],[1/4,3/8,1/4],[1/8,1/4,1/8]]);
I1:=diag(1,1,1);
```

$$A := \begin{bmatrix} \frac{5}{8} & \frac{3}{8} & \frac{5}{8} \\ \frac{1}{4} & \frac{3}{8} & \frac{1}{4} \\ \frac{1}{8} & \frac{1}{4} & \frac{1}{8} \end{bmatrix}$$

$$I1 := \begin{bmatrix} 1 & 0 & 0 \\ 0 & 1 & 0 \\ 0 & 0 & 1 \end{bmatrix}$$

```
> B:=evalm(A-I1);
```

$$B := \begin{bmatrix} \frac{-3}{8} & \frac{3}{8} & \frac{5}{8} \\ \frac{1}{4} & \frac{-5}{8} & \frac{1}{4} \\ \frac{1}{8} & \frac{1}{4} & \frac{-7}{8} \end{bmatrix}$$

```
> x:=linsolve(B,[0,0,0]);
```

$$x := \left[\frac{31}{9}\_t_1, \frac{16}{9}\_t_1, \_t_1\right]$$

The components of x must add up to one which gives us $\_t_1$.

```
> w:=solve(x[1]+x[2]+x[3]=1,_t[1]);
```

$$w := \frac{9}{56}$$

```
> S:=map2(subs,_t[1]=9/56,x);
```

$$S := \left[\frac{31}{56}, \frac{2}{7}, \frac{9}{56}\right]$$

Note that the components of the solution do in fact add up to one.

Finally we mention the determinant. Only square matrices can have a determinant. If A is an m by n matrix then it is square if m = n otherwise it is not. The maple command for finding the determinant is **det.** It is known that the matrix equation Ax = b, where A is an n by n matrix, has a unique solution if $\det(A) \neq 0$. This is equivalent to saying that A is non singular if $\det(A) \neq 0$.

Exercises for chapter 3

Below solve the matrix equation Ax = b by

a). reducing the augmented matrix to quasi upper triangular form using the **pivot** command. If the system is inconsistent point this fact out.

b). Using **gausselim** followed by **backsub.**

c). Using **linsove.**

In each case check your answer by computing Ax – b to see if it is zero.

1). $A = \begin{bmatrix} 1 & -1 & 0 \\ 3 & -5 & 7 \\ 6 & 0 & 2 \end{bmatrix}$, $b^T = [1,-4,2]$.

2). $A=\begin{bmatrix} 7 & -3 & 12 \\ 5 & -5 & 7 \\ 6 & -4 & 2 \end{bmatrix}$, $b^T=[2,5,0]$.

3). $A=\begin{bmatrix} 6 & 6 & 12 \\ -2 & -5 & 8 \\ 5 & -4 & 2 \end{bmatrix}$, $b^T=[1,-1,2]$.

4). $A=\begin{bmatrix} 2 & -3 & 6 \\ 8 & -2 & 0 \\ 8 & 4 & 5 \end{bmatrix}$, $b^T=[0,-4,8]$.

5). For each of the matrices above and using the **pivot** and **mulrow** command find the row echelon form and the reduced row echelon form,

Each of the matrix equations Ax = b has an infinite number of solutions, For each problem find at least 5 particular solutions by assigning values to the parameters.

6). $A=\begin{bmatrix} 1 & -1 & 3 \\ 2 & -2 & 6 \\ 4 & -4 & 12 \end{bmatrix}$, $b^T=[2,4,8]$.

7). $A=\begin{bmatrix} 1 & -1 & 3 \\ 3 & -5 & 7 \\ 4 & -6 & 10 \end{bmatrix}$, $b^T=[2,4,6]$.

8). For each of the following matrices find the inverse by:

a). Forming [A|I] and using **gaussjord,** and **submatrix** to get the inverse.

b). Using the **inverse** command to obtain the inverse.

A). A1 := matrix([[-58, -90, -53, -69], [-84, 46, 59, -56], [-83, -91, 92, -93], [91, -54, 10, -77]]).

B). A2 := matrix([[-63, -90, 61, -3], [-82, 16, -40, 21], [-94, -98, 75, 39], [95, -68, 98, -36]]).

C). A3 := matrix([[-95, 8, 92, 8], [-95, -18, 44, 66], [-62, 40, 68, -67], [68, -65, 43, 6]]).

D). A4 := matrix([[39, -67, 8, 20], [93, 45, 81, 8], [-44, -80, -5, 23], [34, -81, -95, 63]]).

9). Find the determinant for each of the four matrices in problem 8.

10). For each of the matrices below find the rank and the solution to the matrix equation $Ax = 0$ using the **linsolve** command.

A). B1 := matrix([[39, -67, -28, -28], [93, 45, 138, 138], [-44, -80, -124, -124], [34, -81, -47, -47]])

B). B2 := matrix([[39, -67, 8, 413], [93, 45, 81, -39], [-44, -80, -5, 312], [34, -81, -95, 473]])

11). For convenience, each set of column vectors below is written as a set of row vectors. Determine whether or not each set is linearly independent.

a). S = {[1,2,3],[4,5,6],[7,8,9]}.

b). S = {[1,-2,3],[0,1,4],[1,5,0]}.

c). S = {[1,2,0,-4],[3,4,0,-2]}.

d). S = {[1,0,2,-4],[4,3,2,0],[5,3,5,-4]}.

Project 1.

In this project we will be looking at the set of polynomials of degree three or less $P4=\{a_0+a_1x+a_2x^2+a_3x^3 : a_k \text{ is a scalar}, k=1,2,3,4\}$. We shall identify polynomials of this type with the vector $[a_0,a_1,a_3,a_4]$. This will make it easy to add polynomials and to multiply them by scalars (real numbers). To see this let

$p1=a_0+a_1x+a_2x^2+a_3x^3,\quad p2=b_0+b_1x+b_2x^2+b_3x^3$ which are associated with the vectors. $v1=[a_0,a_1,a_2,a_3],\quad v2=[b_0,b_1,b_2,b_3]$. Thus we have

1). $\beta v1=[\beta a_0,\beta a_1,\beta a_2,\beta a_3]$ while $\beta p1=\beta a_0+\beta a_1x+\beta a_2x^2+\beta a_3x^3$.

2). $v1+v2=[a_0+b_0,a_1+b_1,a_2+b_2,a_3+b_3]$ and
$p1+p2=a_0+b_0+(a_1+b\ )x+(a_2+b_2)x^2+(a_3+b_3)x^3$.

Note that we can work with either the polynomials or there equivalent vectors. It makes no difference since the vector form of the polynomial contains the same information as does the original polynomial.

Definition A subset S of P4 is a subspace if

1). $p1\in S\Rightarrow \beta p1\in S$.

2). $p1,p2\in S\Rightarrow p1+p2\in S$.

Definition Let P be the set of all polynomials with real coefficients and S a subset of P. Let $T=\{p1,p2,\ldots,p_m\}$ be a subset of S then T is a basis for S if

1). The polynomials (vectors) in T are linearly independent that is $\sum_{k=1}^{m}\beta_k p_k=0$ for all x implies that $\beta_k=0\ \forall k=1,2,\ldots,m$.

2). The set T spans S that is given p in S there exists scalars β_k such that $p=\sum_{k=1}^{m}\beta_k p_k$.

In testing either 1 or 2 in the above definition we can test either the vectors associated with the polynomial or the polynomial itself. Answer the following questions.

1). Show that P4 is a subspace of the set of all polynomials.

2). Find 2 distinct subspaces of P4 that are not the whole space. Show that they satisfy the definition for subspace.

3). Determine if the set
$T=\{v1=[1,-1,0-2],v2=[0,1,-1,3],v3=[1,2,-1,1],v4=[0,0,3,5]\}$ is a basis for P4.

4). Let $S = \{[a_0, 0, a_2, 0] : a_0 + a_2 x^2 \text{ where } a_0, a_2 \text{ are scalars}\}$. Determine if S is a subspace of P4.

5). Let $T = \{v1 = [1,1,0,0], v2 = [0,0,1,1]\}$ and describe the subspace of P4 for which T is a basis. That is indicate what a typical polynomial in the span of T would look like.

6). Let S be the subset of P4 such that if $p1 = a_0 + a_1 x + a_2 x^2 + a_3 x^3 \in S$ then $\sum_{k=0}^{3} a_k = 0$.

Determine if S is a subspace of P4.

Project 2.

Let $R^n = \{x = (x_1, x_2, \ldots, x_n) : x_k$ is a scalar $k = 1, 2, \ldots, n\}$ be the set of all n vectors. On R^n. We have the following operations:

1). $\beta x = (\beta x_1, \beta x_2, \ldots, \beta x_n)$.

2). $x + y = (x_1, x_2, \ldots, x_n) + (y_1, y_2, \ldots, y_n) = (x_1 + y_1, x_2 + y_2, \ldots, x_n + y_n)$.

We shall be interested in subspaces of R^n and so we define them as follows

Definition A subset S of R^n is a subspace if

1). $p1 \in S \Rightarrow \beta p1 \in S$.

2). $p1, p2 \in S \Rightarrow p1 + p2 \in S$.

Given a subspace one thing we are always interested in is a basis for the subspace. A basis for a subspace space is defined as follows:

Definition Let S be a subset of R^n. Let $T = \{v1, v2, \ldots, v_m\}$ be a subset of S then T is a basis for S if

1). The vectors in T are linearly independent that is $\sum_{k=1}^{m} \beta_k p_k = 0$ for all x implies that $\beta_k = 0\ \forall k = 1, 2, \ldots, m$.

2). The set T spans S that is given p in S there exists scalars β_k such that $p = \sum_{k=1}^{m} \beta_k p_k$.

Answer the following questions.

1). Let $S = \{x = (z, 3z, 3z, 4z) : z \text{ is a scalar}\}$ show that S either is a subspace of R^n or is not.

2). Find a basis for the subspace S of problem (1).

3). Let $S = \{v = (3x_1, 0, -2x_2, 0)\}$ show that S is a subspace of R^n.

4). Find a basis for the set S of problem (3).

5). Let $T = \{v_1 = (1, 2, -1, 0), v_2 = (0, -1, 1, 1), v_3 = (0, 1, -4, 0)\}$ and show that the vectors in the set T are linearly independent.

6). Let $Q = (u = (1, 3, -11, 2), u_2 = (1, 1, 2, 1), u_3 = (1, 3, -2, -1), u_4 = (3, -1, 2, 1))$. Determine which of these vectors, if any, are in the span of T (from problem 5). For those vectors that are in the span of T find constants β_k such that $u = \beta_1 v_1 + \beta_2 v_2 + \beta_3 v_3$.

7). Show that $T = \{v_1 = (1,-1,2), v_2 = (2,-1,0), v_3 = (0,-1,1)\}$ is a basis for R^3 and express $u = (1,2,3)$ as a linear combination of the basis elements.

Project 3

This project involves Markov chains. Consider the case of a taxi driver who alternates between the airport (A) and four hotels which we will label as h1, h2, h3 and h4. We will Assign the probabilities

$$A = \begin{array}{c} \\ \end{array}\begin{matrix} A & h1 & h2 & h3 & h4 \end{matrix}$$

$$A = \begin{bmatrix} 0 & 1/2 & 1/2 & 1/2 & 5/8 \\ 1/4 & 0 & 1/8 & 1/8 & 1/8 \\ 1/4 & 1/8 & 0 & 1/8 & 1/8 \\ 1/8 & 1/8 & 1/8 & 0 & 1/8 \\ 3/8 & 2/8 & 2/8 & 2/8 & 0 \end{bmatrix} \begin{matrix} A \\ h1 \\ h2 \\ h3 \\ h4 \end{matrix}$$

We interpret the Markov matrix as follows A[3,4] = 1/8 which means that if the taxi driver is currently at hotel h3 then the probability that he will next drive to hotel h2 is 1/8. A[1,5] = 3/8 means that if the taxi driver is at the airport now then the probability that he will drive next to hotel h4 is 3/8. From the matrix we can see that hotel h4 is the more popular hotel.

Answer the following questions:

1). If the taxi driver is currently at the airport now, what are the probabilities for where he will be next?

2). Given $x_0^T = [1,0,0,0,0]$ compute the sequence $x_1 = Ax_0, x_2 = Ax_1, \ldots, x_{10} = Ax_9$. Is there any pattern emerging?

3). Use a loop to compute the sequence in problem (3) out to x_{100}. Print out this last vector along with its predecessor. Has convergence taken place yet?

4). What does the answer to problem (3) tell us? That is where is the taxi driver most likely to be spending most of his time?

5). Does this problem have a fixed point, that is, is there a vector x such that Ax = x ? Either find the fixed point x or show that it does not exist.

Project 4

In Figure 1 below we have a fluid flow problem. The constants give us the water flow into the node in terms of cubic centimeters per second. What we want to do is to determine the flow in the various pipes. The solution here may not be unique. The governing rule for our equations is

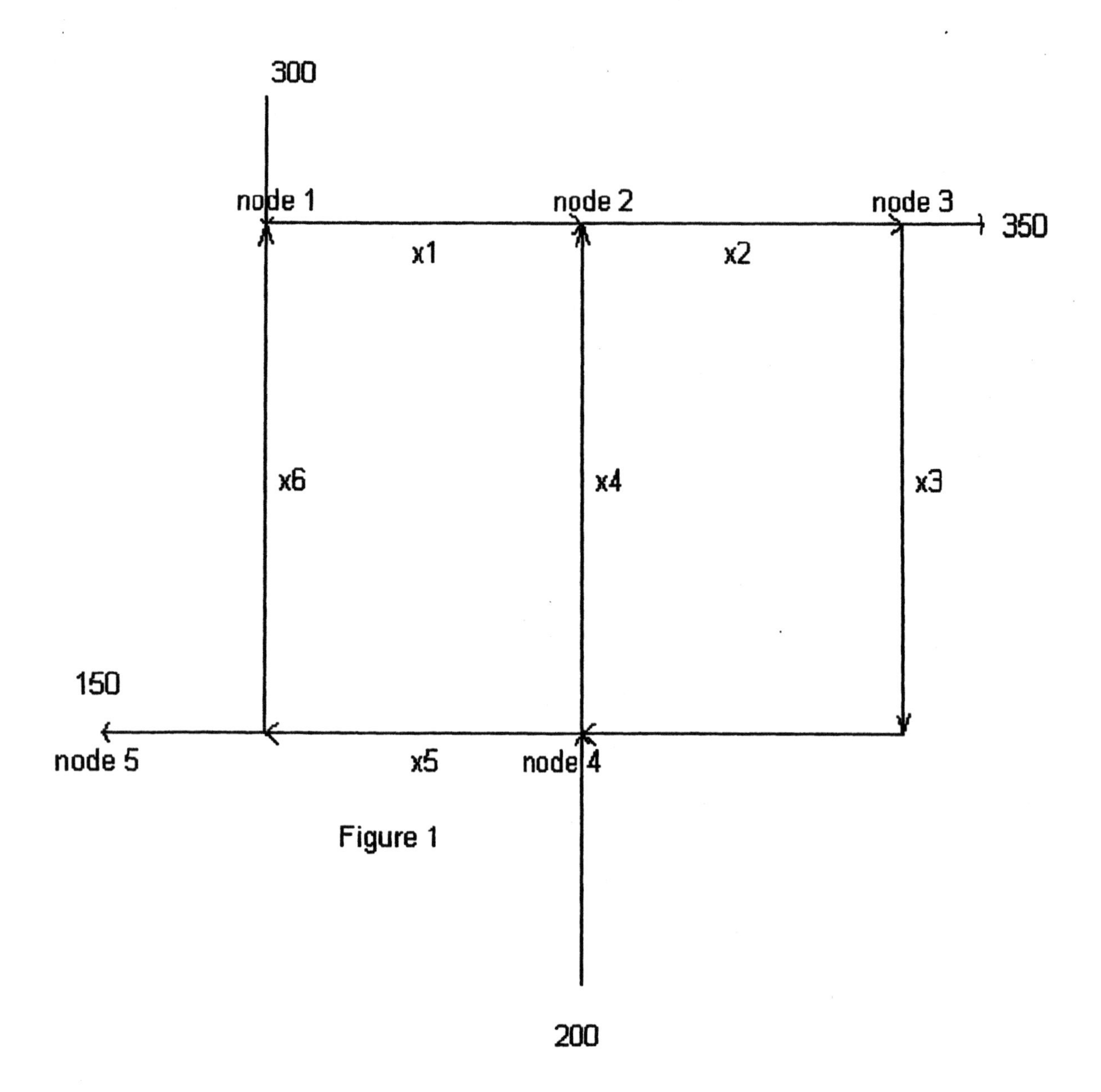

Figure 1

Node Rule
The sum of the flows into the node must equal the sum of the flows out of the node.

This rule simply says that water cannot disappear at a node or alternatively be created at a node.

1). Write out the matrix equation Ax = b for the system shown in Figure 1.

2). Solve the system of equations generated in problem (1). Is the solution unique?

3). Is it possible to find a solution where x1, x2, x4, x5, and x6 are all positive?

4). Change the direction of the input at node 5 and resolve.

5). From the original model change the direction of the inputs at both node 3 and 5.

6). Is it possible to get an all negative solution which means that the direction of all of the x vectors will be changed?

7). Going back to the original model in problem (1) suppose that the pipe with flow x5 is closed for repair. What will be the solution for the flow in the remaining pipes?

Chapter 3.1
Solving systems of equations using the LinearAlgebra package

1. Equation Reduction and the Augmented Matrix

In chapter 2 we saw that the system of equations

$$a_{11}x_1 + a_{12}x_2 + a_{13}x_3 = b_1$$
$$a_{21}x_1 + a_{22}x_2 + a_{23}x_3 = b_2$$
$$a_{31}x_1 + a_{32}x_2 + a_{33}x_3 = b_3$$

could be written as the matrix equation Ax = b where $A = \begin{bmatrix} a_{11} & a_{12} & a_{13} \\ a_{21} & a_{22} & a_{23} \\ a_{31} & a_{32} & a_{33} \end{bmatrix}$,

$b = [b_1, b_2, b_3]^T$ and $x = [x_1, x_2, x_3]^T$. There are three operations that we can perform on the system of equations without changing the solution. They are:

1). Interchange two equations.

2). Multiply an equation by a non zero scalar (number).

3). Multiply equation j by a non zero scalar and add the result to equation m where j does not equal m.

Now consider the augmented matrix $[A \mid b] = \begin{bmatrix} a_{11} & a_{12} & a_{13} & b_1 \\ a_{21} & a_{22} & a_{23} & b_2 \\ a_{31} & a_{32} & a_{33} & b_3 \end{bmatrix}$. This matrix contains all the essential information in the system of equations. The three operations given above are now performed on the rows of [A|b]. The object is to reduce this augmented matrix to either row echelon form or reduced row echelon form. Once we have one of these two forms we can easily solve the system of equations.

In our reductions below we shall use quasi echelon forms in the sense that we will not insist on having a one in the usual position but will be happy with any non zero value.

2. Reduction to Echelon Form

The maple commands that will be useful for this section are **Pivot, GaussianElimination, ReducedRowEchelonForm,** and **LinearSlove.** Maple has a single command that will solve the system of equations Ax = b. The maple command is **LinearSolve.** Nearly all Linear Algebra texts begin by solving these equations using the three rules stated at the beginning of section 1. For this reason then we shall first solve the system of equations by the rather long and tedious method that is shown in your text. We shall use the commands **RandonMatrix** and **RandomVector** to generate random matrices and random vectors.

As an example we shall solve a matrix equation using the Pivot command.

```
> restart; with(LinearAlgebra):
> A:=RandomMatrix(4,4); b:=RandomVector(4);
```

$$A := \begin{bmatrix} -41 & -34 & -56 & 62 \\ 20 & -62 & -8 & -79 \\ -7 & -90 & -50 & -71 \\ 16 & -21 & 30 & 28 \end{bmatrix}$$

$$b := \begin{bmatrix} -65 \\ 5 \\ 66 \\ -36 \end{bmatrix}$$

```
> A1:=<A|b>;
```

$$A1 := \begin{bmatrix} -41 & -34 & -56 & 62 & -65 \\ 20 & -62 & -8 & -79 & 5 \\ -7 & -90 & -50 & -71 & 66 \\ 16 & -21 & 30 & 28 & -36 \end{bmatrix}$$

```
> B1:=Pivot(A1,1,1);
```

$$B1 := \begin{bmatrix} -41 & -34 & -56 & 62 & -65 \\ 0 & \frac{-3222}{41} & \frac{-1448}{41} & \frac{-1999}{41} & \frac{-1095}{41} \\ 0 & \frac{-3452}{41} & \frac{-1658}{41} & \frac{-3345}{41} & \frac{3161}{41} \\ 0 & \frac{-1405}{41} & \frac{334}{41} & \frac{2140}{41} & \frac{-2516}{41} \end{bmatrix}$$

```
> B2:=Pivot(B1,2,2,2..4);
```

$$B2 := \begin{bmatrix} -41 & -34 & -56 & 62 & -65 \\ 0 & \frac{-3222}{41} & \frac{-1448}{41} & \frac{-1999}{41} & \frac{-1095}{41} \\ 0 & 0 & \frac{-4190}{1611} & \frac{-47281}{1611} & \frac{56767}{537} \\ 0 & 0 & \frac{37934}{1611} & \frac{236675}{3222} & \frac{-53399}{1074} \end{bmatrix}$$

```
> B3:=Pivot(B2,3,3,3..4);
```

$$B3 := \begin{bmatrix} -41 & -34 & -56 & 62 & -65 \\ 0 & \frac{-3222}{41} & \frac{-1448}{41} & \frac{-1999}{41} & \frac{-1095}{41} \\ 0 & 0 & \frac{-4190}{1611} & \frac{-47281}{1611} & \frac{56767}{537} \\ 0 & 0 & 0 & \frac{-805539}{4190} & \frac{3801729}{4190} \end{bmatrix}$$

```
> x:=BackwardSubstitute(B3);
```

$$x := \begin{bmatrix} \frac{-5581445}{268513} \\ \frac{-644346}{268513} \\ \frac{3386273}{268513} \\ \frac{-1267243}{268513} \end{bmatrix}$$

Thus the **Pivot** command has rather quickly reduced the augmented matrix to upper triangular (quasi row echelon) form. It can now be easily solved with the maple command **BackwardSubstitute.** This gives


```
> x:=BackwardSubstitute(B3);
```

$$x := \begin{bmatrix} \frac{-5581445}{268513} \\ \frac{-644346}{268513} \\ \frac{3386273}{268513} \\ \frac{-1267243}{268513} \end{bmatrix}$$

The last parameter in the **Pivot** command controls where the zeroes will go. If we use an alternate form of the command we will get the quasi reduced row echelon form as shown in

```
> C1:=Pivot(A1,1,1);
```

$$C1 := \begin{bmatrix} -41 & -34 & -56 & 62 & -65 \\ 0 & \frac{-3222}{41} & \frac{-1448}{41} & \frac{-1999}{41} & \frac{-1095}{41} \\ 0 & \frac{-3452}{41} & \frac{-1658}{41} & \frac{-3345}{41} & \frac{3161}{41} \\ 0 & \frac{-1405}{41} & \frac{334}{41} & \frac{2140}{41} & \frac{-2516}{41} \end{bmatrix}$$

```
> C2:=Pivot(C1,2,2);
```

$$C2 := \begin{bmatrix} -41 & 0 & \frac{-65600}{1611} & \frac{133865}{1611} & \frac{-28700}{537} \\ 0 & \frac{-3222}{41} & \frac{-1448}{41} & \frac{-1999}{41} & \frac{-1095}{41} \\ 0 & 0 & \frac{-4190}{1611} & \frac{-47281}{1611} & \frac{56767}{537} \\ 0 & 0 & \frac{37934}{1611} & \frac{236675}{3222} & \frac{-53399}{1074} \end{bmatrix}$$

```
> C3:=Pivot(C2,3,3);
```

$$C3 := \begin{bmatrix} -41 & 0 & 0 & \frac{227345}{419} & \frac{-715860}{419} \\ 0 & \frac{-3222}{41} & 0 & \frac{30043539}{85895} & \frac{-125591949}{85895} \\ 0 & 0 & \frac{-4190}{1611} & \frac{-47281}{1611} & \frac{56767}{537} \\ 0 & 0 & 0 & \frac{-805539}{4190} & \frac{3801729}{4190} \end{bmatrix}$$

```
> C4:=Pivot(C3,4,4);
```

$$C4 := \begin{bmatrix} -41 & 0 & 0 & 0 & \frac{228839245}{268513} \\ 0 & \frac{-3222}{41} & 0 & 0 & \frac{2076082812}{11009033} \\ 0 & 0 & \frac{-4190}{1611} & 0 & \frac{-14188483870}{432574443} \\ 0 & 0 & 0 & \frac{-805539}{4190} & \frac{3801729}{4190} \end{bmatrix}$$

In this case we can easily calculate the answer by using **BackwardSubstitue** command to get the solution. We have

```
> x:=BackwardSubstitute(C4);
```

$$x := \begin{bmatrix} \frac{-5581445}{268513} \\ \frac{-644346}{268513} \\ \frac{3386273}{268513} \\ \frac{-1267243}{268513} \end{bmatrix}$$

Finally we look at the maple commands **GaussianElimination** and **ReducedRowEchelonForm.** These commands are the fastest way of arriving at the above quasi echelon forms.

```
> D1:=GaussianElimination(A1);
```

$$D1 := \begin{bmatrix} -41 & -34 & -56 & 62 & -65 \\ 0 & \frac{-3222}{41} & \frac{-1448}{41} & \frac{-1999}{41} & \frac{-1095}{41} \\ 0 & 0 & \frac{-4190}{1611} & \frac{-47281}{1611} & \frac{56767}{537} \\ 0 & 0 & 0 & \frac{-805539}{4190} & \frac{3801729}{4190} \end{bmatrix}$$

```
> D2:=ReducedRowEchelonForm(A1);
```

$$D2 := \begin{bmatrix} 1 & 0 & 0 & 0 & \frac{-5581445}{268513} \\ 0 & 1 & 0 & 0 & \frac{-644346}{268513} \\ 0 & 0 & 1 & 0 & \frac{3386273}{268513} \\ 0 & 0 & 0 & 1 & \frac{-1267243}{268513} \end{bmatrix}$$

We can obtain the solution from D1 (or D2) by applying the **BackwardSubstitute** command. When a system of equations has a unique solution such as in the above case we say that A is nonsingular. When there are infinitely many solutions we say that A is singular. When there are no solutions we say that the system of equations is inconsistent.

3. Singular systems

We can determine that a system of equations is singular from its echelon form. If it is singular then its echelon form will contain one or more rows of zeroes. Consider the following example of an inconsistent system of equations:

```
> restart; with(LinearAlgebra):
> A:=Matrix(4,4,(i,j)->i+j); b1:=Vector([3,-2,0,1]);
```

$$A := \begin{bmatrix} 2 & 3 & 4 & 5 \\ 3 & 4 & 5 & 6 \\ 4 & 5 & 6 & 7 \\ 5 & 6 & 7 & 8 \end{bmatrix}$$

$$b1 := \begin{bmatrix} 3 \\ -2 \\ 0 \\ 1 \end{bmatrix}$$

```
> A1:=<A|b1>;
```

$$A1 := \begin{bmatrix} 2 & 3 & 4 & 5 & 3 \\ 3 & 4 & 5 & 6 & -2 \\ 4 & 5 & 6 & 7 & 0 \\ 5 & 6 & 7 & 8 & 1 \end{bmatrix}$$

```
> A2:=ReducedRowEchelonForm(A1);
```

$$A2 := \begin{bmatrix} 1 & 0 & -1 & -2 & 0 \\ 0 & 1 & 2 & 3 & 0 \\ 0 & 0 & 0 & 0 & 1 \\ 0 & 0 & 0 & 0 & 0 \end{bmatrix}$$

It is easy to see that this system of equations is inconsistent because of row 3. The equation corresponding to row 3 is $0x_1 + 0x_2 + 0x_3 + 0x_4 = 1$ for which there is obviously no solution. We can modify the vector b to get an infinite number of solutions as in:

```
> b2:=Vector([1,1,1,1]);
```

$$b2 := \begin{bmatrix} 1 \\ 1 \\ 1 \\ 1 \end{bmatrix}$$

```
> A3:=<A|b2>;
```

$$A3 := \begin{bmatrix} 2 & 3 & 4 & 5 & 1 \\ 3 & 4 & 5 & 6 & 1 \\ 4 & 5 & 6 & 7 & 1 \\ 5 & 6 & 7 & 8 & 1 \end{bmatrix}$$

```
> A4:=ReducedRowEchelonForm(A3);
```

$$A4 := \begin{bmatrix} 1 & 0 & -1 & -2 & -1 \\ 0 & 1 & 2 & 3 & 1 \\ 0 & 0 & 0 & 0 & 0 \\ 0 & 0 & 0 & 0 & 0 \end{bmatrix}$$

```
> x:=BackwardSubstitute(A4);
```

$$x := \begin{bmatrix} -1 + \_t_2 + 2\_t_1 \\ 1 - 2\_t_2 - 3\_t_1 \\ \_t_2 \\ \_t_1 \end{bmatrix}$$

The occurrence of the parameters $\_t_1$, $\_t_2$ indicate that there are an infinite number of solutions. We can get particular solutions by assigning particular values to the parameters as in

```
> x1:=map2(subs,{_t[1]=0,_t[2]=0},x);
```

$$x1 := \begin{bmatrix} -1 \\ 1 \\ 0 \\ 0 \end{bmatrix}$$

```
> x2:=map2(subs,{_t[1]=1,_t[2]=0},x);
```

$$x2 := \begin{bmatrix} 1 \\ -2 \\ 0 \\ 1 \end{bmatrix}$$

```
> x3:=map2(subs,{_t[1]=1,_t[2]=1},x);
```

$$x3 := \begin{bmatrix} 2 \\ -4 \\ 1 \\ 1 \end{bmatrix}$$

4. The inverse Matrix and the LU decomposition

B is the matrix inverse to the matrix A if AB = I and BA = I where I is the identity matrix. The inverse matrix exists if A is nonsingular that is if the equation Ax = b has a unique solution. When it exists we denote the inverse matrix by the symbol A^{-1}.

Suppose that we have a series of matrix equations to solve such as $Ax_1 = b_1, Ax_2 = b_2, \cdots, Ax_m = b_m$ we can solve all of these simultaneously as follows. Let X be the matrix with columns $x_1, x_2, \ldots, x_m$ and B be the matrix with columns $b_1, b_2, \ldots, b_m$ then the individual problems are contained in the single matrix equation AX = B. A solution can be found by reducing the augmented matrix [A|B] to reduced row echelon form.

To find the inverse matrix, when it exists, we need to solve the matrix equation AX = I. We can start with the augmented matrix [A|I] and proceed to reduce this to reduced row echelon form which will be $\left[I \mid A^{-1}\right]$. For example

```
> restart; with(LinearAlgebra):
> A:=RandomMatrix(3,3); I1:=DiagonalMatrix(<1,1,1>);
```

$$A := \begin{bmatrix} 97 & 55 & 13 \\ -82 & 68 & -65 \\ -66 & 26 & 5 \end{bmatrix}$$

$$I1 := \begin{bmatrix} 1 & 0 & 0 \\ 0 & 1 & 0 \\ 0 & 0 & 1 \end{bmatrix}$$

```
> B:=<A|I1>;
```

$$B := \begin{bmatrix} 97 & 55 & 13 & 1 & 0 & 0 \\ -82 & 68 & -65 & 0 & 1 & 0 \\ -66 & 26 & 5 & 0 & 0 & 1 \end{bmatrix}$$

```
> B1:=ReducedRowEchelonForm(B);
```

$$B1 := \begin{bmatrix} 1 & 0 & 0 & \frac{145}{34717} & \frac{9}{69434} & \frac{-637}{69434} \\ 0 & 1 & 0 & \frac{2350}{243019} & \frac{1343}{486038} & \frac{5239}{486038} \\ 0 & 0 & 1 & \frac{1178}{243019} & \frac{-3076}{243019} & \frac{5553}{243019} \end{bmatrix}$$

To pick out the inverse we can use maple's **SubMatrix** command. As parameters we provide the underlying matrix name, a list of the rows to keep and finally a list of the columns to keep. This will define the sub matrix that we want to extract.

```
> Ainv:=SubMatrix(B1, [1,2,3], [4,5,6]);
```

$$Ainv := \begin{bmatrix} \frac{145}{34717} & \frac{9}{69434} & \frac{-637}{69434} \\ \frac{2350}{243019} & \frac{1343}{486038} & \frac{5239}{486038} \\ \frac{1178}{243019} & \frac{-3076}{243019} & \frac{5553}{243019} \end{bmatrix}$$

In practice, of course, we use the maple function **MatrixInverse** to get the inverse matrix as in

```
> B3:=MatrixInverse(A);
```

$$B3 := \begin{bmatrix} \frac{145}{34717} & \frac{9}{69434} & \frac{-637}{69434} \\ \frac{2350}{243019} & \frac{1343}{486038} & \frac{5239}{486038} \\ \frac{1178}{243019} & \frac{-3076}{243019} & \frac{5553}{243019} \end{bmatrix}$$

The inverse matrix can be used to solve the matrix equation Ax = b as

$$Ax = b \Rightarrow A^{-1}Ax = A^{-1}b$$
$$Ix = A^{-1}b.$$
$$x = A^{-1}b$$

Although this is theoretically acceptable it is not used in solving large systems of equations. Rather various transformations are used instead. One of these is the LU decomposition.

P is a permutation matrix if it is derived from the identity matrix by permuting the rows of the identity matrix. A permutation matrix P satisfies the relationship PP = I. The maple command **LUDecomposition** will give the following decomposition of a matrix A, A = PLU. Here P is a permutation matrix, L is a unit lower triangular matrix, that is, it has ones along the main diagonal, and U is an upper triangular matrix. Once we have the decomposition we solve Ax = b as PLUx = b and since PP = I this gives LUx = Pb. We then solve this problem as

$$Ly = Pb$$
$$Ux = y$$

This may seem to be a very cumbersome and tedious way to solve the problem but actually there are very good numerical reasons for using methods of this type.

Example 4.1

Using the LU decomposition solve the matrix equation Ax = b where

$$A = \begin{bmatrix} 1 & -5 & 3 \\ 2 & 0 & -7 \\ 15 & -3 & 4 \end{bmatrix} \text{ and } b^T = [1,2,3].$$

Solution

```
>A:=Matrix([[1,-5,3],[2,0,-7],[15,-3,4]]);
```

$$A := \begin{bmatrix} 1 & -5 & 3 \\ 2 & 0 & -7 \\ 15 & -3 & 4 \end{bmatrix}$$

```
>b:=Vector([1,2,3]);
```

$$b := \begin{bmatrix} 1 \\ 2 \\ 3 \end{bmatrix}$$

```
> (P,L,U):=LUDecomposition(A);
```

$$P, L, U := \begin{bmatrix} 1 & 0 & 0 \\ 0 & 1 & 0 \\ 0 & 0 & 1 \end{bmatrix}, \begin{bmatrix} 1 & 0 & 0 \\ 2 & 1 & 0 \\ 15 & \frac{36}{5} & 1 \end{bmatrix}, \begin{bmatrix} 1 & -5 & 3 \\ 0 & 10 & -13 \\ 0 & 0 & \frac{263}{5} \end{bmatrix}$$

```
>
> y:=ForwardSubstitute(<L|b>);
```

$$y := \begin{bmatrix} 1 \\ 0 \\ -12 \end{bmatrix}$$

```
> x:=BackwardSubstitute(<U|y>);
```

$$x := \begin{bmatrix} \frac{53}{263} \\ \frac{-78}{263} \\ \frac{-60}{263} \end{bmatrix}$$

Just as a check on our work we have

```
> check:=evalm(A.x-b);
```

$$check := [0, 0, 0]$$

The zero result shows that we have correctly found the solution to the original problem. For an example where the matrix P is not the identity matrix consider

Example 4.2

Using the LU decomposition solve the matrix equation Ax = b where $A = \begin{bmatrix} 1 & -1 & 3 \\ 2 & -2 & 0 \\ 5 & -3 & 1 \end{bmatrix}$ and $b^T = [1, 2, 3]$.

Solution

```
> A:=Matrix([[1,-1,3],[2,-2,0],[5,-3,1]]);
```

$$A := \begin{bmatrix} 1 & -1 & 3 \\ 2 & -2 & 0 \\ 5 & -3 & 1 \end{bmatrix}$$

```
> b:=Vector([1,2,3]);
```

$$b := \begin{bmatrix} 1 \\ 2 \\ 3 \end{bmatrix}$$

```
> (P,L,U):=LUDecomposition(A);
```

$$P, L, U := \begin{bmatrix} 1 & 0 & 0 \\ 0 & 0 & 1 \\ 0 & 1 & 0 \end{bmatrix}, \begin{bmatrix} 1 & 0 & 0 \\ 5 & 1 & 0 \\ 2 & 0 & 1 \end{bmatrix}, \begin{bmatrix} 1 & -1 & 3 \\ 0 & 2 & -14 \\ 0 & 0 & -6 \end{bmatrix}$$

```
>y:=ForwardSubstitute(<L|P.b>);
```

$$y := \begin{bmatrix} 1 \\ -2 \\ 0 \end{bmatrix}$$

```
>x:=BackwardSubstitute(U,y);
```

$$x := \begin{bmatrix} 0 \\ -1 \\ 0 \end{bmatrix}$$

```
>check:=evalm(A.x-b);
```

$$check := [0, 0, 0]$$

5. Linear Independence, Rank, and Determinant

A set of vectors $S = \{v_1, v_2, \ldots, v_m\}$ is said to be linearly independent if $\sum_{k=1}^{m} c_k v_k = 0$ can hold only if each of the constants $c_k = 0$. Otherwise they are linearly dependent. This idea is not unique to vectors, for instance we can let the v's be matrices of the same size. Linear independence is an idea that applies to a wide variety of situations, In Linear Algebra we are primarily concerned with vectors and shall stick with them. Let A be the matrix whose first column is v_1, whose second column is v_2 etc. Then the vectors in the set S are linearly independent if and only if the equation Ac = 0 has the unique solution c = 0. For sets of vectors this is an easy test.

Example 5.1

Determine if the set S of vectors given below is linearly independent or not.
$S = \{v_1 = [1,2,-1,3], v_2 = [2,0,3,-1], v_3 = [3,2,2,2]\}$.

Solution

```
>A:=Matrix([[1,2,3],[2,0,2],[-1,3,2],[3,-1,2]]);
```

$$A := \begin{bmatrix} 1 & 2 & 3 \\ 2 & 0 & 2 \\ -1 & 3 & 2 \\ 3 & -1 & 2 \end{bmatrix}$$

```
>b:=Vector([0,0,0,0]);
```

$$b := \begin{bmatrix} 0 \\ 0 \\ 0 \\ 0 \end{bmatrix}$$

```
> c:=LinearSolve(A,b);
```

$$c := \begin{bmatrix} -\_t0_3 \\ -\_t0_3 \\ \_t0_3 \end{bmatrix}$$

From the answer for c it is clear that there are an infinite number of solutions to the equation, in particular [-1,-1,1], and hence the vectors are linearly dependent. Another important property of a matrix is its rank. The rank of a matrix A is the number of linearly independent columns in A. The number of linearly independent columns is always the same as the number of linearly independent rows in A.

In the above example we could have answered the question by finding the rank of A, it is 2. The fact that the rank is less than the number of columns tells us that the columns are linearly dependent. In an m by n matrix A the columns are linearly independent if and only if the rank is n. The maple command to determine the rank is **Rank.**

Example 5.2

In this example we shall revisit an example of a Markov chain from chapter 2. A Markov matrix is one in which every entry satisfies $0 \le a_{ij} \le 1$ and where the sum of the entries in each column is one. For this example we shall set C = cloudy, R = rainy, and S = sunny. We assume that C, R, and S are the only weather possibilities. Set

$$\begin{array}{c} \text{today} \\ \begin{array}{ccc} S & C & R \end{array} \end{array}$$

$$A = \begin{bmatrix} \frac{5}{8} & \frac{3}{8} & \frac{5}{8} \\ \frac{1}{4} & \frac{3}{8} & \frac{1}{4} \\ \frac{1}{8} & \frac{1}{4} & \frac{1}{8} \end{bmatrix} \begin{array}{l} S \\ C \\ R \end{array} \text{ tomorrow}$$

We interpret the matrix A as follows. If it is sunny today then the probability that it will be cloudy tomorrow is A[2,1] = 1/4 or 0.25. If it is cloudy today then the probability that it will be sunny tomorrow is A[1,2] = 3/8.We enter the data into the matrix in rational form because this will give more accurate results when we compute powers of A.

Now let S0 = [1/2,1/4,1/4] be the probabilities of S, C, and R for today's weather. The probability for tomorrows weather then is S1 = A(S0). The probability for the weather the day after tomorrow will be S2 = A(S1) = AA(S0) etc. The question that we want to answer is does the sequence of vectors S0, S1, S2,... approach a limiting value? In the example in chapter 2 we saw that the answer was probably yes. How can we find the limiting vector? If we denote the limiting vector by S then we have AS = S. This is the equation whose solution, if it exists, is the limiting vector. Solutions of this type are called fixed points. We can solve the matrix equation by noting that AS – IS = 0 where I is the identity matrix. Thus we must solve (A-I)S = 0. Before solving the equation we need to note that we are looking for a nontrivial solution S. Furthermore the entries in S must all add up to 1. With that in mind then we can now solve the equation

```
> A:=Matrix([[5/8,3/8,5/8],[1/4,3/8,1/4],[1/8,1/4,1/8]]);
I1:=IdentityMatrix(3);
```

$$A := \begin{bmatrix} \frac{5}{8} & \frac{3}{8} & \frac{5}{8} \\ \frac{1}{4} & \frac{3}{8} & \frac{1}{4} \\ \frac{1}{8} & \frac{1}{4} & \frac{1}{8} \end{bmatrix}$$

$$I1 := \begin{bmatrix} 1 & 0 & 0 \\ 0 & 1 & 0 \\ 0 & 0 & 1 \end{bmatrix}$$

```
> B:=A-I1; zero:=Vector(3,0):
```

$$B := \begin{bmatrix} \frac{-3}{8} & \frac{3}{8} & \frac{5}{8} \\ \frac{1}{4} & \frac{-5}{8} & \frac{1}{4} \\ \frac{1}{8} & \frac{1}{4} & \frac{-7}{8} \end{bmatrix}$$

```
> x:=LinearSolve(B,zero);
```

$$x := \begin{bmatrix} \frac{31}{9}\_t0_3 \\ \frac{16}{9}\_t0_3 \\ \_t0_3 \end{bmatrix}$$

Now we need to select the parameter so that the sum of the components of x is 1.

```
> w:=solve(x[1]+x[2]+x[3]=1,_t0[3]);
```

$$w := \frac{9}{56}$$

```
> S:=map2(subs,_t0[3]=9/56,x);
```

$$S := \begin{bmatrix} \frac{31}{56} \\ \frac{2}{7} \\ \frac{9}{56} \end{bmatrix}$$

Finally we mention the determinant. Only square matrices can have a determinant. If A is an m by n matrix then it is square if m = n otherwise it is not. The maple command for finding the determinant is **Determinant.** It is known that the matrix equation Ax = b, where A is an n by n matrix, has a unique solution if $\text{Determinant}(A) \neq 0$. This is equivalent to saying that A is non singular if $\text{Determinant}(A) \neq 0$.

Exercises for chapter 3

Below solve the matrix equation Ax = b by

a). reducing the augmented matrix to quasi upper triangular form using the **Pivot** command. If the system is inconsistent point this fact out.

b). Using **GaussianElimination** followed by **BackwardSubstitute.**

c). Using **LinearSolve.**

In each case check your answer by computing Ax – b to see if it is zero.

1). $A = \begin{bmatrix} 1 & -1 & 0 \\ 3 & -5 & 7 \\ 6 & 0 & 2 \end{bmatrix}$, $b^T = [1,-4,2]$.

2). $A = \begin{bmatrix} 7 & -3 & 12 \\ 5 & -5 & 7 \\ 6 & -4 & 2 \end{bmatrix}$, $b^T = [2,5,0]$.

3). $A = \begin{bmatrix} 6 & 6 & 12 \\ -2 & -5 & 8 \\ 5 & -4 & 2 \end{bmatrix}$, $b^T = [1,-1,2]$.

4). $A = \begin{bmatrix} 2 & -3 & 6 \\ 8 & -2 & 0 \\ 8 & 4 & 5 \end{bmatrix}$, $b^T = [0,-4,8]$.

5). For each of the matrices above and using the **Pivot** and **ReducedRowEchelonForm** command find the row echelon form and the reduced row echelon form,

Each of the matrix equations Ax = b has an infinite number of solutions, For each problem find at least 5 particular solutions by assigning values to the parameters.

6). $A=\begin{bmatrix} 1 & -1 & 3 \\ 2 & -2 & 6 \\ 4 & -4 & 12 \end{bmatrix}$, $b^T=[2,4,8]$.

7). $A=\begin{bmatrix} 1 & -1 & 3 \\ 3 & -5 & 7 \\ 4 & -6 & 10 \end{bmatrix}$, $b^T=[2,4,6]$.

8). For each of the following matrices find the inverse by:

a). Forming [A|I] and using **ReducedRowEchelonForm,** and **SubMatrix** to get the inverse.

b). Using the Matrix**Inverse** command to obtain the inverse.

A). A1 := Matrix([[-58, -90, -53, -69], [-84, 46, 59, -56], [-83, -91, 92, -93], [91, -54, 10, -77]]).

B). A2 := Matrix([[-63, -90, 61, -3], [-82, 16, -40, 21], [-94, -98, 75, 39], [95, -68, 98, -36]]).

C). A3 := Matrix([[-95, 8, 92, 8], [-95, -18, 44, 66], [-62, 40, 68, -67], [68, -65, 43, 6]]).

D). A4 := Matrix([[39, -67, 8, 20], [93, 45, 81, 8], [-44, -80, -5, 23], [34, -81, -95, 63]]).

9). Find the determinant for each of the four matrices in problem 8.

10). For each of the matrices below find the rank and the solution to the matrix equation Ax = 0 using the **LinearSolve** command.

A). B1 := Matrix([[39, -67, -28, -28], [93, 45, 138, 138], [-44, -80, -124, -124], [34, -81, -47, -47]])

B). B2 := Matrix([[39, -67, 8, 413], [93, 45, 81, -39], [-44, -80, -5, 312], [34, -81, -95, 473]])

11). For convenience, each set of column vectors below is written as a set of row vectors. Determine whether or not each set is linearly independent.

a). S = {[1,2,3],[4,5,6],[7,8,9]}.

b). $S = \{[1,-2,3],[0,1,4],[1,5,0]\}$.

c). $S = \{[1,2,0,-4],[3,4,0,-2]\}$.

d). $S = \{[1,0,2,-4],[4,3,2,0],[5,3,5,-4]\}$.

Chapter 4.0
Eigenvalues, Eigenvectors, and Orthogonal Vectors

Let A be an n by n matrix and x a vector and let us ask the question when will Ax be parallel to x. That is when will Ax be in either the same or exactly opposite direction to x? That question leads us to the following definition.

Definition 4.0

Let A be an n by n matrix, x a nonzero vector and λ a scalar then λ is an eigenvalue of A and x its corresponding non zero eigenvector if Ax = λx.

Thus if Ax is parallel to the vector x then x is an eigenvector of A. Eigenvalues can be complex valued. Some properties of eigenvalues that we will be using in this chapter are:

Properties

1). If A is a symmetric matrix then all of its eigenvalues are real valued.

2). If A is a symmetric matrix then eigenvectors corresponding to distinct eigenvalues are orthogonal.

3). Eigenvectors are not unique for if Ax = λx and c is any nonzero scalar then A(cx) = λ(cx). Even if we require x to be of unit length they are still not unique since there will be two of them.

Now we turn our attention to finding the eigenvalues of a matrix A.

Finding Eigenvalues and Eigenvectors

We wish to find the eigenvalues and eigenvectors of an n by n matrix A. That is all the solutions to the matrix equation Ax = λx or Ax = λIx where I is the n by n identity matrix. This matrix equation can be written as Ax - λIx = 0 or (A - λI)x = 0. This last equation is a homogeneous matrix equation. We know from the theory of linear algebra that a homogeneous matrix equation has a non zero solution if and only if the determinant of the underlying matrix is zero. This means that the eigenvalues of A are the solutions of the characteristic equation that is det(A - λI) = 0.

The characteristic equation is a polynomial of degree n in λ. As such it has n roots some of which may be real, complex, or repeated. If p(x) is a polynomial of degree n and a particular root, say r, appears m times in the list of all n roots then r is said to have multiplicity m. We can find the characteristic polynomial of a matrix A by using the maple function charpoly.

```
> restart; with(linalg):
Warning, the protected names norm and trace have been redefined and
unprotected
```

```
>A:=randmatrix(5,5,symmetric);
```

$$A := \begin{bmatrix} -85 & -55 & -35 & 79 & 57 \\ -55 & -37 & 97 & 56 & -59 \\ -35 & 97 & 50 & 49 & 45 \\ 79 & 56 & 49 & 63 & -8 \\ 57 & -59 & 45 & -8 & -93 \end{bmatrix}$$

```
>p:=charpoly(A,lambda);
```

$p := \lambda^5 + 102\,\lambda^4 - 40910\,\lambda^3 - 2736974\,\lambda^2 + 370491474\,\lambda + 311683396$

How can we find the roots of this polynomial? We can use maple's solve **or** fsolve **command. Using these commands we have:**

```
>w:=fsolve(p=0,lambda);
```

$w := -184.7839911, -158.2792595, -0.8361696501, 77.54355473, 164.3558656$

To find the individual eigenvectors we proceed as follows.

```
>I1:=diag(1,1,1,1,1): zero:=vector([0,0,0,0,0]):
```

```
>x:=linsolve(A-w[1]*I1,zero);
```

$$x := [0., -0., 0., 0., -0.]$$

We use I1 instead of I because maple uses I in complex numbers. W[1] = -184.7839911 is the first of the eigenvalues listed in w. The solution for x is zero indicating that there are no non zero solutions. This does not mean that w[1] fails to be an eigenvalue, it simply indicates the difficulty of computing with floating point numbers. W[1] is a ten digit approximation **to an eigenvalue which may take an infinite number of digits to specify accurately.**

How can we get around this problem?

2. The Maple commands eigenvals and eigenvectors

The maple command eigenvectors **will compute numerical approximations to the eigenvalues and eigenvectors provided that at least one of the entries in the matrix A is in floating point. To satisfy this condition we will change the entry A[1,1] to floating point.**

```
>A[1,1]:=-85.0;
```

$$A_{1,1} := -85.0$$

Now that we have met that requirement we should be able to get all of the information that we want from:

```
> w1:=eigenvectors(A);
```

wl := [164.3558656, 1, {
[-0.02369669116, 0.4885164988, 0.6586761703, 0.5714389127, -0.01983350357]}
], [-184.7839911, 1,
{[-0.5869341484, 0.1844663604, -0.3412393331, 0.2345794147, 0.6708268649]}]
, [77.54355480, 1,
{[0.5632053855, -0.3103231987, -0.3254182354, 0.6699328940, 0.1783033783]}]
, [-158.2792591, 1,
{[0.5390568122, 0.6987657451, -0.2213436721, -0.3103192517, 0.2754144606]}]
, [-0.8361696431, 1,
{[-0.2171732444, 0.3778223616, -0.5429653491, 0.2707802524, -0.6647952537]}
]

Starting at the topmost left we have listed the eigenvalue 164.3558656, next the number one indicates that the multiplicity of this eigenvalue is one. This information is followed by a set giving all of the associated eigenvectors, in this case one. If the multiplicity had been 3 then the set would have contained 3 eigenvectors (since A is symmetric). Thus
λ = 164.3558656 and

x = [-0.02369669116,0.4885164988,0.6586761703,0.5714389127,-0.01983350357].

These numbers are long and tedious to type. There is a maple function op **that can help us. To get the pieces that we want. We may have to experiment around as in:**

```
> nops(w1[4]); # how many operands in w1[4]?
```

3

```
> lambda_1:=op(1,w1[4]);
```

lambda_1 := -158.2792591

```
> m_1:=op(2,w1[4]);
```

m_1 := 1

```
> x_1:=op(3,w1[4]);
```

x_1 :=
{[0.5390568122, 0.6987657451, -0.2213436721, -0.3103192517, 0.2754144606]}

The first line tells us that there are 3 operands in the fourth component of w1. The second line accesses the eigenvalue while the line after that tells us that the multiplicity is 1. We accessed this for illustrative purposes only since we would normally not need it. Finally the solution for x_1 is a set containing only one element

that is one eigenvector. We access the elements of a set with the same notation that we use to access elements of a vector. Finally we use these commands to pick out the eigenvalue and eigenvector in w1[5] .

```
> lambda_1:=op(1,w1[5]);
```

$$lambda\_1 := -0.8361696431$$

```
> x_1:=op(3,w1[5])[1];
```

$x\_1 := [-0.2171732444, 0.3778223616, -0.5429653491, 0.2707802524, -0.6647952537]$

Now lets look at a case where the eigenvalues are integers so that we can solve exactly for the eigenvectors.

```
> A:=matrix([[-9,21,7,9,13],[-4,5,4,4,4],[18,-39,-8,-14,-22],[-26,42,14,22,26],[8,6,-8,-8,0]]);
```

$$A := \begin{bmatrix} -9 & 21 & 7 & 9 & 13 \\ -4 & 5 & 4 & 4 & 4 \\ 18 & -39 & -8 & -14 & -22 \\ -26 & 42 & 14 & 22 & 26 \\ 8 & 6 & -8 & -8 & 0 \end{bmatrix}$$

For illustrative purposes only we will show how to solve for the eigenvalues and eigenvectors without using the maple functions eigenvals **and** eigenvectors. **These two commands are by far the most efficient way of coming up with the desired information. The more primitive methods that we are going to employ are for the illustration of certain maple commands only.**

```
> p:=charpoly(A,lambda);
```

$$p := \lambda^5 - 10\,\lambda^4 + 3\,\lambda^3 + 118\,\lambda^2 - 88\,\lambda - 192$$

```
> w:=solve(p=0,lambda);
```

$$w := -1, 2, -3, 4, 8$$

```
> I1:=diag(1,1,1,1,1): zero:=vector([0,0,0,0,0]):
```

Again we use I1 rather than I since I is a reserved word in maple. Now we can solve for the eigenvector associated with the eigenvalue w[1] = -1.

```
> x:=linsolve(A-w[1]*I1,zero);
```

$$x := [\_t_1, 2\,\_t_1, 0, 2\,\_t_1, -4\,\_t_1]$$

```
> x1:=map2(subs,_t[1]=1,x); # x1 is an eigenvector for w[1].
```

$$x1 := [1, 2, 0, 2, -4]$$

In this way we can find all of the eigenvectors. We will limit ourselves to the additional eigenvalue
W[5] = 8.

```
>x:=linsolve(A-w[5]*I1,zero);
```

$$x := [\_t_1, 0, -2\_t_1, 2\_t_1, \_t_1]$$

```
>x5:=map2(subs,_t[1]=1,x); # x5 is an eigenvector for w[5].
```

$$x5 := [1, 0, -2, 2, 1]$$

Notice that we can solve for the eigenvectors using this method only because we have the exact eigenvalue and not just a 10 digit approximation.

3. Diagonalization and powers

Suppose that $Ax_i = \lambda_i x_i \; i = 1,2,3,\ldots,n$ **and let** $P = [x_1, x_2, \ldots, x_n]$ **be the matrix whose columns are the eigenvectors** $x_i \; i = 1,2,\ldots,n$**. Furthermore let** $D = diag(\lambda_1, \lambda_2, \ldots, \lambda_n)$ **be a diagonal matrix with the eigenvalues of the matrix A on the main diagonal. Then we have**
AP = PD or $A = PDP^{-1}$**. If this can be done then we say that the matrix A can be diagonalized or that A is similar to a diagonal matrix D.**

Not every matrix can be diagonalized. Every symmetric matrix can be diagonalized. **What happens in the case where a matrix cannot be diagonalized? It can be shown that every matrix A is similar to its Jordan canonical form. The Jordan canonical form looks like**

$$J = \begin{bmatrix} J1 & 0 & 0 \\ . & . & . \\ 0 & 0 & Jl \end{bmatrix}$$

in block form where each block Jk looks like

$$Jk = \begin{bmatrix} \lambda & 1 & 0 & 0 & 0 \\ 0 & \lambda & 1 & 0 & 0 \\ . & . & . & . & . \\ 0 & 0 & 0 & \lambda & 1 \\ 0 & 0 & 0 & 0 & \lambda \end{bmatrix}$$

where we have ones on the super diagonal and the associated eigenvalue on the main diagonal. The sizes of the blocks vary and can be 1 by 1 in many cases. There is a maple function jordan **which will give us the decomposition. Consider the following example.**

```
>A:=matrix([[-15,-11,1,1,5,6,6,7],[24,16,0,-2,-10,-12,-12,-
14],[34,24,-1,-2,-15,-18,-18,-21],[11,7,1,-3,-5,-6,-6,-
7],[-42,-24,1,3,18,18,18,21],[29,13,1,-3,-15,-14,-18,-
21],[-24,-13,0,2,10,12,16,15],[24,12,0,-2,-10,-12,-12,-
10]]);
```

$$A := \begin{bmatrix} -15 & -11 & 1 & 1 & 5 & 6 & 6 & 7 \\ 24 & 16 & 0 & -2 & -10 & -12 & -12 & -14 \\ 34 & 24 & -1 & -2 & -15 & -18 & -18 & -21 \\ 11 & 7 & 1 & -3 & -5 & -6 & -6 & -7 \\ -42 & -24 & 1 & 3 & 18 & 18 & 18 & 21 \\ 29 & 13 & 1 & -3 & -15 & -14 & -18 & -21 \\ -24 & -13 & 0 & 2 & 10 & 12 & 16 & 15 \\ 24 & 12 & 0 & -2 & -10 & -12 & -12 & -10 \end{bmatrix}$$

Now to find its Jordan decomposition we use.

```
>J:=jordan(A,'P'):   # Here we have AP = PJ.
>print(J,P);
```

$$\begin{bmatrix} 3 & 0 & 0 & 0 & 0 & 0 & 0 & 0 \\ 0 & 4 & 1 & 0 & 0 & 0 & 0 & 0 \\ 0 & 0 & 4 & 0 & 0 & 0 & 0 & 0 \\ 0 & 0 & 0 & -2 & 1 & 0 & 0 & 0 \\ 0 & 0 & 0 & 0 & -2 & 1 & 0 & 0 \\ 0 & 0 & 0 & 0 & 0 & -2 & 0 & 0 \\ 0 & 0 & 0 & 0 & 0 & 0 & -2 & 0 \\ 0 & 0 & 0 & 0 & 0 & 0 & 0 & 4 \end{bmatrix}, \begin{bmatrix} -1 & 1 & \frac{-3}{2} & \frac{-1}{6} & \frac{-13}{36} & \frac{-277}{216} & \frac{-31}{36} & 0 \\ 2 & -2 & 3 & 0 & \frac{1}{3} & \frac{61}{18} & 2 & 0 \\ 3 & -3 & \frac{9}{2} & \frac{-1}{6} & \frac{5}{36} & \frac{1145}{216} & \frac{113}{36} & 0 \\ 1 & -1 & \frac{3}{2} & \frac{-1}{6} & \frac{-1}{36} & \frac{419}{216} & \frac{41}{36} & 0 \\ -4 & 3 & \frac{-9}{2} & \frac{-1}{6} & \frac{-25}{36} & \frac{-1009}{216} & \frac{-103}{36} & 0 \\ 3 & -3 & 1 & \frac{-1}{6} & \frac{11}{36} & \frac{1187}{216} & \frac{113}{36} & -1 \\ -2 & 3 & -2 & 0 & \frac{-1}{3} & \frac{-61}{18} & -2 & 1 \\ 2 & -2 & 4 & 0 & \frac{1}{3} & \frac{61}{18} & 2 & 0 \end{bmatrix}$$

Just for a quick check on our work we have.

```
> evalm(A&*P-P&*J);
```

$$\begin{bmatrix} 0 & 0 & 0 & 0 & 0 & 0 & 0 & 0 \\ 0 & 0 & 0 & 0 & 0 & 0 & 0 & 0 \\ 0 & 0 & 0 & 0 & 0 & 0 & 0 & 0 \\ 0 & 0 & 0 & 0 & 0 & 0 & 0 & 0 \\ 0 & 0 & 0 & 0 & 0 & 0 & 0 & 0 \\ 0 & 0 & 0 & 0 & 0 & 0 & 0 & 0 \\ 0 & 0 & 0 & 0 & 0 & 0 & 0 & 0 \\ 0 & 0 & 0 & 0 & 0 & 0 & 0 & 0 \end{bmatrix}$$

which shows that AP = PJ or alternatively $A = PJP^{-1}$.

It should be noted that the Jordan canonical form can only be computed accurately when all of the eigenvalues are known exactly. It can be shown that if we only have approximations to the eigenvalues then we cannot compute the Jordan canonical form with enough accuracy to be useful.

If A can be diagonalized then we can make use of this in computing powers of A. To see this note that

$$A = PDP^{-1} \Rightarrow$$
$$A^2 = (PDP^{-1})(PDP^{-1}) = PD^2P^{-1}.$$

In general we can use mathematical induction to show that $A^n = PD^nP^{-1}$. As an example we have:

```
> restart; with(linalg):
Warning, the protected names norm and trace have been redefined and
unprotected
```

```
> A:=matrix([[-9,21,7,9,13],[-4,5,4,4,4],[18,-39,-8,-14,-
22],[-26,42,14,22,26],[8,6,-8,-8,0]]);
```

$$A := \begin{bmatrix} -9 & 21 & 7 & 9 & 13 \\ -4 & 5 & 4 & 4 & 4 \\ 18 & -39 & -8 & -14 & -22 \\ -26 & 42 & 14 & 22 & 26 \\ 8 & 6 & -8 & -8 & 0 \end{bmatrix}$$

```
> w:=eigenvectors(A);
```

$$w := \left[4, 1, \{\left[\frac{-1}{2}, 0, 1, \frac{-3}{2}, 0\right]\}\right], [8, 1, \{[1, 0, -2, 2, 1]\}], [-3, 1, \{[1, 1, -1, 2, -2]\}], [2, 1, \{[1, 0, -1, 2, 0]\}], [-1, 1, \{[1, 2, 0, 2, -4]\}]$$

```
> D1:=diag(4,8,-3,2,-1):
>
P:=augment(op(3,w[1])[1],op(3,w[2])[1],op(3,w[3])[1],op(3,w
[4])[1],op(3,w[5])[1]);
```

$$P := \begin{bmatrix} \frac{-1}{2} & 1 & 1 & 1 & 1 \\ 0 & 0 & 1 & 0 & 2 \\ 1 & -2 & -1 & -1 & 0 \\ \frac{-3}{2} & 2 & 2 & 2 & 2 \\ 0 & 1 & -2 & 0 & -4 \end{bmatrix}$$

Now that we have AP = PD1 we shall use our method to compute $A^{20} = PD1^{20}P^{-1}$.

```
> A_20:=evalm(P&*D1^20&*inverse(P));
```
$A\_20 :=$

[-2192047589599 , 2305842998752292175 , -6972520225 , 1092538058975 , 1152921497633278175]
[6973568800 , -10460353199 , -6973568800 , -6973568800 , -6973568800]
[4391070845150 , -4611686007965986125 , 6972520226 , -2192049686750 , -2305843002240125150]
[-6583118434750 , 4611685997504584350 , -13945040450 , 3284587745726 , 2305842995266556350]
[-13947137600 , 2305843030134400350 , 13947137600 , 13947137600 , 1152921518553984576]

Exactly the same technique can be used when the diagonal matrix D is replaced by the Jordan canonical form J. In this case computing J^n is not as efficient as computing D^n because J is upper triangular and not diagonal. Hence rather quickly the powers will fill up the zero positions in the upper triangular portion of J^n.

4. Positive Definite Matrices

Definition 4.1
The n by n symmetric matrix A is said to be positive definite if $x^T Ax > 0\ \forall x \neq 0$.

We shall need the following property of symmetric matrices.

Property
If A is an n by n symmetric matrix then there exists an orthonormal matrix P such that
$P^T P = I$ **and a diagonal matrix D such that AP = PD. Thus D contains the eigenvalues and P the eigenvectors of A.**

Theorem 4.1
The n by n symmetric matrix A is positive definite if and only if all of the eigenvalues of A are positive.

This is an easy test for us to apply using maple. To see this let us generate 3 random 5 by 5 matrices and check to see if they are positive definite or not.

```
> restart; with(linalg):
Warning, the protected names norm and trace have been redefined and
unprotected
```

```
> A:=randmatrix(5,5,symmetric);
```

$$A := \begin{bmatrix} -85 & -55 & -35 & 79 & 57 \\ -55 & -37 & 97 & 56 & -59 \\ -35 & 97 & 50 & 49 & 45 \\ 79 & 56 & 49 & 63 & -8 \\ 57 & -59 & 45 & -8 & -93 \end{bmatrix}$$

```
> A[1,1]:=-85.0;   w:=eigenvalues(A); # Not positive definite
```

$$A_{1,1} := -85.0$$

$$w := -184.7839911, -158.2792595, -0.8361696501, 77.54355473, 164.3558656$$

```
> A:=randmatrix(5,5,symmetric);
```

$$A := \begin{bmatrix} 92 & 43 & 77 & -5 & -12 \\ 43 & -62 & 66 & 99 & -18 \\ 77 & 66 & 54 & -61 & 31 \\ -5 & 99 & -61 & -50 & -26 \\ -12 & -18 & 31 & -26 & -62 \end{bmatrix}$$

```
> A[1,1]:=92.0;   w:=eigenvalues(A); # Not positive definite
```

$$A_{1,1} := 92.0$$

$$w := -188.9263358, -79.05938399, 0.8465155062, 62.11545810, 177.0237462$$

```
> A:=randmatrix(5,5,symmetric);
```

$$A := \begin{bmatrix} 1 & -47 & -47 & -58 & 94 \\ -47 & -91 & -61 & -90 & 83 \\ -47 & -61 & 41 & 53 & -86 \\ -58 & -90 & 53 & -1 & 23 \\ 94 & 83 & -86 & 23 & -84 \end{bmatrix}$$

```
> A[1,1]:=1.0;   w:=eigenvalues(A); # Not positive definite
```

$$A_{1,1} := 1.0$$

$$w := -254.3259190, -88.74407294, -6.741946544, 26.49968519, 189.3122533$$

None of the three randomly generated matrices had all positive eigenvalues and so none of them are positive definite.

For symmetric matrices the fact that the eigenvector matrix P can be chosen to satisfy $P^T P = I$ means that the columns of P are orthonormal, that is they are orthogonal and have a length of one. Orthogonal matrices have very nice properties. One of the questions that we now want to study is this:

Question

Given a set of n linearly independent vectors S find an orthonormal set of n vectors Q such that span(S) = span(Q).

We can then replace the basis S of span(S) with the basis Q which is orthonormal. As we will see later this will be of great advantage especially when it comes to expanding a given element of span(Q) as a linear combination of the basis elements.

5. Orthonormal vectors and the Gram-Schmidt Process

The Gram-Schmidt process is an algorithm that will create the set Q from the set S. In fact once the algorithm is complete we will have A decomposed as A = QR where $Q^T Q = I$. To see how the process works set

$S = \{v_1, v_2, \ldots, v_n\}$ and $Q = \{u_1, u_2, \ldots, u_n\}$. **Then set**

$$u_1 = v_1 / \|v_1\|_2$$

$$w = v_k - \sum_{j=1}^{k-1} r_{jk} u_j \quad k = 2,3,\ldots,n$$

$$u_m^T w = 0 = u_m^T v_k - r_{mk} \Rightarrow r_{mk} = u_m^T v_k \quad m = 1,2,\ldots,k-1$$

$$u_k = w / \|w\|_2$$

$$r_{kk} = \|w\|_2$$

To implement this in maple w need to write it as an algorithm that can be implemented. We assume that the vectors v are available as well as an appropriately sized matrix R which has been filled with zeros.

Algorithm (Gram-Schmidt)

1). $u_1 = v_1 / \|v_1\|_2$

2). $R[1,1] = \|v_1\|_2$

3). For k from 2 to n

4). For j from 1 to k-1

5) $R[j,k] = u_j^T v_k$

6). End for j

7). $w = v_k - \sum_{j-1}^{k-1} R[j,k] u_j$

8). $u_k = w / \|w\|_2$

9). $R[k,k]=\|w\|_2$

10). End for k

We shall write up our algorithm as a maple procedure and apply it to an example.

```
> restart; with(linalg):
Warning, the protected names norm and trace have been redefined and
unprotected
```

Gram-Schmidt procedure.

```
> GS:=proc(S::matrix,R::matrix) local j,k,m,n,w,w1,Q,u,z;
> w:=col(S,1);
> R[1,1]:=evalf(norm(w,2));
> w1:=map(evalf,evalm(w/R[1,1]));
> Q:=augment(w1);
> n:=coldim(S);
> for k from 2 to n do
>  for j from 1 to k-1 do
>   R[j,k]:=innerprod(col(Q,j),col(S,k));
>   u:=vector(k-1,i->R[i,k]);
>  od:
>  w:=evalm(col(S,k)-Q&*u);
>  z:=map(evalf,evalm(w/norm(w,2)));
>  R[k,k]:=norm(w,2);
>  Q:=augment(Q,z);
> od:
> evalm(Q);
>  end:
```

The above procedure would be used on many occasions and so we really should write it up as a text file which we can read into any program on an as needed basis. Also the text file form is much easier to modify when adding or deleting features. Now lets apply the procedure to a matrix.

```
> m:=5: S:=randmatrix(m,m);  rank(S);
```

$$S := \begin{bmatrix} -61 & 41 & -58 & -90 & 53 \\ -1 & 94 & 83 & -86 & 23 \\ -84 & 19 & -50 & 88 & -53 \\ 85 & 49 & 78 & 17 & 72 \\ -99 & -85 & -86 & 30 & 80 \end{bmatrix}$$

5

```
> R:=matrix(m,m,0):
> Q:=GS(S,R); print(R);
```

$$Q := \begin{bmatrix} -0.3658273702, & 0.4431117203, & -0.6410153129, & -0.4307974996, & 0.27080476 \\ -0.005997170003, & 0.7034136620, & 0.6619719776, & -0.2587511727, & -0.0037663696 \\ -0.5037622803, & 0.3307730200, & -0.04618326655, & 0.7937242722, & -0.068422656 \\ 0.5097594503, & 0.1742007604, & -0.05290057225, & 0.3150641580, & 0.779573924 \\ -0.5937198303, & -0.4112232087, & 0.3820485629, & -0.1348988565, & 0.56056527 \end{bmatrix}$$

$$\begin{bmatrix} 166.7453148 & 50.31025915 & 136.7294789 & -20.03654498 & -3.62229067 \\ 0 & 134.0629622 & 65.09705843 & -80.64088737 & 1.77706340 \\ 0 & 0 & 57.44930443 & 7.259807796 & 10.45430086 \\ 0 & 0 & 0 & 132.1812367 & -58.95822002 \\ 0 & 0 & 0 & 0 & 118.8669716 \end{bmatrix}$$

We should check two things to make sure that our procedure is operating correctly. First we check that $Q^T Q = I$ and then we check that A – QR is the zero matrix.

```
> print(evalm(transpose(Q)&*Q));
```

$$\begin{bmatrix} 1.000000000 & 0.2\ 10^{-9} & 0.9\ 10^{-9} & 0.9\ 10^{-10} & -0.2\ 10^{-9} \\ 0.2\ 10^{-9} & 0.9999999997 & 0.6\ 10^{-9} & 0.12\ 10^{-9} & -0.1\ 10^{-9} \\ 0.9\ 10^{-9} & 0.6\ 10^{-9} & 0.9999999995 & 0.36\ 10^{-9} & 0.3\ 10^{-9} \\ 0.9\ 10^{-10} & 0.12\ 10^{-9} & 0.36\ 10^{-9} & 1.000000001 & 0.28\ 10^{-9} \\ -0.2\ 10^{-9} & -0.1\ 10^{-9} & 0.3\ 10^{-9} & 0.28\ 10^{-9} & 1.000000000 \end{bmatrix}$$

```
> print(map(simplify,evalm(S-Q&*R)));
```

$$\begin{bmatrix} 0.1000000000\ 10^{-7}, & -0.1000000000\ 10^{-7}, & 0., & 0., & 0. \\ 0., & 0., & 0., & 0.1000000000\ 10^{-7}, & 0. \\ 0.1000000000\ 10^{-7}, & -0.1000000000\ 10^{-7}, & 0., & 0., & 0. \\ -0.1000000000\ 10^{-7}, & -0.1000000000\ 10^{-7}, & 0., & 0., & 0.1000000000\ 10^{-7} \\ 0.1000000000\ 10^{-7}, & 0., & 0., & 0., & 0. \end{bmatrix}$$

Again since we are using floating point numbers, which may be approximations only and not exact, we do not get the identity matrix as we should. Since the off diagonal elements are very small (in ten digit arithmetic) we may treat them as zeroes. Also A – QR is not the zero matrix but the entries in the difference are either zero or small and hence we may interpret them as zero.

6. QR factorization

Maple also has a function QRdecomp **which decomposes A as A = QR where $QQ^T = I$ and R is upper triangular. You should read the help page for** QRdecomp. **The resulting Q will supply an orthonormal basis for the column space of A and if A is singular an orthonormal basis for the null space of A if we find .** $A^T = QR$

Example 6.1

Given the matrix

$$A = \begin{bmatrix} 1 & 0 & 0 & -1 & 0 & -1 \\ 2 & 1 & -1 & -2 & 1 & -1 \\ 5 & 4 & -3 & -6 & 4 & -1 \\ -5 & 0 & 4 & 1 & 0 & 5 \\ 9 & -8 & 15 & -16 & -8 & -17 \\ -9 & -5 & -4 & 18 & -5 & 4 \end{bmatrix}$$

1). Perform the QR decomposition and print out R

2). Show that the first three columns of Q are a basis for the column space of A.

3). Show that if we find the QR decomposition of the transpose of A then the last three columns of Q will be a basis for the null space (kernel) of A.

•

Solution

Question # (1).

```
> restart; with(linalg):
Warning, the protected names norm and trace have been redefined and
unprotected
```

```
> A:=matrix([[1,0,0,-1,0,-1],[2,1,-1,-2,1,-1],[5,4,-3,-6,4,-
1],[-5,0,4,1,0,5],[9,-8,15,-16,-8,-17],[-9,-5,-4,18,-
5,4]]);
```

$$A := \begin{bmatrix} 1 & 0 & 0 & -1 & 0 & -1 \\ 2 & 1 & -1 & -2 & 1 & -1 \\ 5 & 4 & -3 & -6 & 4 & -1 \\ -5 & 0 & 4 & 1 & 0 & 5 \\ 9 & -8 & 15 & -16 & -8 & -17 \\ -9 & -5 & -4 & 18 & -5 & 4 \end{bmatrix}$$

```
> rank(A);
```

$$3$$

```
> w:=colspace(A);
```

$w := \{[1, 0, 0, 7, 46, -26], [0, 1, 0, -16, -36, 31], [0, 0, 1, 4, 7, -9]\}$

```
> C:=augment(w[1],w[2],w[3]);
```

$$C := \begin{bmatrix} 1 & 0 & 0 \\ 0 & 1 & 0 \\ 0 & 0 & 1 \\ 7 & -16 & 4 \\ 46 & -36 & 7 \\ -26 & 31 & -9 \end{bmatrix}$$

```
> R:=QRdecomp(A,Q='Q');
```

$$R := \begin{bmatrix} \sqrt{217} & -\frac{5\sqrt{217}}{217} & \frac{134\sqrt{217}}{217} & -\frac{346\sqrt{217}}{217} & -\frac{5\sqrt{217}}{217} & -\frac{222\sqrt{217}}{217} \\ 0 & \frac{3\sqrt{554001}}{217} & -\frac{1037\sqrt{554001}}{72261} & \frac{38\sqrt{554001}}{72261} & \frac{3\sqrt{554001}}{217} & \frac{3\sqrt{554001}}{217} \\ 0 & 0 & \frac{\sqrt{7779990}}{333} & -\frac{\sqrt{7779990}}{333} & 0 & 0 \\ 0 & 0 & 0 & 0 & 0 & 0 \\ 0 & 0 & 0 & 0 & 0 & 0 \\ 0 & 0 & 0 & 0 & 0 & 0 \end{bmatrix}$$

```
> Q1:=augment(col(Q,1),col(Q,2),col(Q,3));
```

$$Q1 := \begin{bmatrix} \frac{\sqrt{217}}{217} & \frac{5\sqrt{554001}}{1662003} & -\frac{593\sqrt{7779990}}{23339970} \\ \frac{2\sqrt{217}}{217} & \frac{227\sqrt{554001}}{1662003} & -\frac{574\sqrt{7779990}}{11669985} \\ \frac{5\sqrt{217}}{217} & \frac{893\sqrt{554001}}{1662003} & -\frac{907\sqrt{7779990}}{11669985} \\ -\frac{5\sqrt{217}}{217} & -\frac{25\sqrt{554001}}{1662003} & \frac{6961\sqrt{7779990}}{23339970} \\ \frac{9\sqrt{217}}{217} & -\frac{1691\sqrt{554001}}{1662003} & \frac{676\sqrt{7779990}}{11669985} \\ -\frac{9\sqrt{217}}{217} & -\frac{1130\sqrt{554001}}{1662003} & -\frac{1922\sqrt{7779990}}{11669985} \end{bmatrix}$$

The rank of the matrix is three indicating that the basis of the column space contains 3 vectors. We gather the basis, returned by the maple function colspace, **up into the matrix C. Now we need to show that if w = Cx then w can be expressed as Q1z for some z showing that the span(C) is contained in the span(Q1). Then we will show that if z = Q1x then there exists a y such that Cy = z showing that span(Q1) is contained in span(C) and hence that the two spans are the same. Thus the first 3 columns of Q constitute a basis for the column space of the matrix A.**

```
> x:=vector(3); w:=evalm(C&*x);
```

$$x := \text{array}(1\,..\,3, [\ ])$$

$$w := [x_1, x_2, x_3, 7\,x_1 - 16\,x_2 + 4\,x_3, 46\,x_1 - 36\,x_2 + 7\,x_3, -26\,x_1 + 31\,x_2 - 9\,x_3]$$

```
> y:=linsolve(Q1,w);
```

$$y := \left[\frac{1}{217}\sqrt{217}\,(614\,x_1 - 521\,x_2 + 129\,x_3), \frac{1}{72261}\sqrt{2553}\,\sqrt{217}\,(2112\,x_1 - 1151\,x_2 + 38\,x_3), \frac{1}{7659}\sqrt{1612070}\,\sqrt{2553}\,(3\,x_1 - 4\,x_2 + x_3)\right]$$

```
> w:=evalm(Q1&*x);
```

$$w := \left[\frac{1}{217}\sqrt{217}\,x_1 + \frac{5}{1662003}\sqrt{554001}\,x_2 - \frac{593}{23339970}\sqrt{7779990}\,x_3, \frac{2}{217}\sqrt{217}\,x_1 + \frac{227}{1662003}\sqrt{554001}\,x_2 - \frac{574}{11669985}\sqrt{7779990}\,x_3, \frac{5}{217}\sqrt{217}\,x_1 + \frac{893}{1662003}\sqrt{554001}\,x_2 - \frac{907}{11669985}\sqrt{7779990}\,x_3, -\frac{5}{217}\sqrt{217}\,x_1 - \frac{25}{1662003}\sqrt{554001}\,x_2 + \frac{6961}{23339970}\sqrt{7779990}\,x_3, \frac{9}{217}\sqrt{217}\,x_1 - \frac{1691}{1662003}\sqrt{554001}\,x_2 + \frac{676}{11669985}\sqrt{7779990}\,x_3, -\frac{9}{217}\sqrt{217}\,x_1 - \frac{1130}{1662003}\sqrt{554001}\,x_2 - \frac{1922}{11669985}\sqrt{7779990}\,x_3\right]$$

```
> b:=linsolve(C,w);
```

$$b := \left[\frac{1}{116489790270}\sqrt{217}\,(536819310\,x_1 + 350450\,\sqrt{2553}\,x_2 - 593\,\sqrt{217}\,\sqrt{1612070}\,\sqrt{2553}\,x_3), \frac{1}{58244895135}\sqrt{217}\,(536819310\,x_1 + 7955215\,\sqrt{2553}\,x_2 - 574\,\sqrt{217}\,\sqrt{1612070}\,\sqrt{2553}\,x_3), \frac{1}{58244895135}\sqrt{217}\,(1342048275\,x_1 + 31295185\,\sqrt{2553}\,x_2 - 907\,\sqrt{217}\,\sqrt{1612070}\,\sqrt{2553}\,x_3)\right]$$

Question #3

```
> w:=kernel(A);
```

$$w := \{[0, -1, 0, 0, 1, 0], [1, -1, 0, 0, 0, 1], [1, 1, 1, 1, 0, 0]\}$$

```
> C:=augment(w[1],w[2],w[3]);
```

$$C := \begin{bmatrix} 1 & 0 & 1 \\ 1 & -1 & -1 \\ 1 & 0 & 0 \\ 1 & 0 & 0 \\ 0 & 1 & 0 \\ 0 & 0 & 1 \end{bmatrix}$$

```
> evalm(A&*C);
```

$$\begin{bmatrix} 0 & 0 & 0 \\ 0 & 0 & 0 \\ 0 & 0 & 0 \\ 0 & 0 & 0 \\ 0 & 0 & 0 \\ 0 & 0 & 0 \end{bmatrix}$$

The kernel **command returns a basis for the null space of A. Now we shall take the last three columns of Q and put them into a matrix Q1. Note that there are three vectors and they are linearly independent because they are orthogonal. To show that it is a basis we need only show that A&*Q1 is the zero matrix.**

```
> R:=QRdecomp(transpose(A),Q='Q'):
> Q1:=augment(col(Q,4),col(Q,5),col(Q,6)):
> evalm(A&*Q1);
```

$$\begin{bmatrix} 0 & 0 & 0 \\ 0 & 0 & 0 \\ 0 & 0 & 0 \\ 0 & 0 & 0 \\ 0 & 0 & 0 \\ 0 & 0 & 0 \end{bmatrix}$$

The A = QR factorization can be used to solve the matrix equation Ax = b. After finding the QR decomposition the matrix equation becomes QRx = b or $Q^T QRx = Q^T b \Rightarrow Rx = Q^T b$. **Since Q is an orthonormal matrix it has nice numerical properties. We should also note that R is an upper triangular matrix. We can take advantage of this in solving the system of equations. Now lets look at some examples.**

```
A:=randmatrix(5,5);   b:=randvector(5);
```

$$A := \begin{bmatrix} 40 & -78 & 62 & 11 & 88 \\ 1 & 30 & 81 & -5 & -28 \\ 4 & -11 & 10 & 57 & -82 \\ -48 & -11 & 38 & -7 & 58 \\ -94 & -68 & 14 & -35 & -14 \end{bmatrix}$$

$$b := [-9, -51, -73, -73, -91]$$

```
> R:=QRdecomp(A,Q='Q');
```

$R :=$

$$\left[\sqrt{12757}, \frac{3786\sqrt{12757}}{12757}, -\frac{539\sqrt{12757}}{12757}, \frac{4289\sqrt{12757}}{12757}, \frac{1696\sqrt{12757}}{12757}\right]$$

$$\left[0, \frac{\sqrt{1745625195078}}{12757}, -\frac{23766524\sqrt{1745625195078}}{872812597539}, -\frac{958650\sqrt{1745625195078}}{290937532513}, -\frac{44594236\sqrt{1745625195078}}{872812597539}\right]$$

$$\left[0, 0, \frac{2\sqrt{12671544928914066999}}{68418327}, \frac{1318110073\sqrt{12671544928914066999}}{4223848309638022333}, \frac{69070791316\sqrt{12671544928914066999}}{12671544928914066999}\right]$$

$$\left[0, 0, 0, \frac{3\sqrt{12222109566182728426083857}}{185206880737}, -\frac{808917326969390\sqrt{12222109566182728426083857}}{36666328698548185278251571}\right]$$

$$\left[0, 0, 0, 0, \frac{206\sqrt{7115907029976132079636359298}}{197974981019283}\right]$$

```
> print(Q);
```

$$\left[\frac{40\sqrt{12757}}{12757}, -\frac{191081\sqrt{1745625195078}}{290937532513}, \frac{370273747\sqrt{12671544928914066999}}{4223848309638022333}, -\frac{405767533610\sqrt{12222109566182728426083857}}{12222109566182728426083857}, \frac{595420\sqrt{7115907029976132079636359298}}{4845567664017820754 17}\right]$$

$$\left[\frac{\sqrt{12757}}{12757}, \frac{63154\sqrt{1745625195078}}{290937532513}, \frac{1041786264\sqrt{12671544928914066999}}{4223848309638022333}, -\frac{937693207858\sqrt{12222109566182728426083857}}{36666328698548185278251571}, -\frac{6420718\sqrt{7115907029976132079636359298}}{1453670299205346226251}\right]$$

$$\left[\frac{4\sqrt{12757}}{12757}, -\frac{155471\sqrt{1745625195078}}{1745625195078}, \frac{203050507\sqrt{12671544928914066999}}{12671544928914066999},\right.$$

$$\frac{3400368837328\sqrt{12222109566182728426083857}}{12222109566182728426083857},$$

$$\left.-\frac{2168029\sqrt{71159070299761320796363592 98}}{96911353280356415083 4}\right]$$

$$\left[-\frac{48\sqrt{12757}}{12757}, \frac{41401\sqrt{1745625195078}}{1745625195078}, \frac{1269135283\sqrt{12671544928914066999}}{12671544928914066999}\right.$$

$$, \frac{1644327833692\sqrt{12222109566182728426083857}}{36666328698548185278251571},$$

$$\left.\frac{28141823\sqrt{7115907029976132079636359298}}{29073405984106924525 02}\right]$$

$$\left[-\frac{94\sqrt{12757}}{12757}, -\frac{255796\sqrt{1745625195078}}{872812597539},\right.$$

$$-\frac{133490246\sqrt{12671544928914066999}}{12671544928914066999},$$

$$-\frac{933543693727\sqrt{12222109566182728426083857}}{36666328698548185278251571},$$

$$\left.-\frac{6631726\sqrt{7115907029976132079636359298}}{1453670299205346226251}\right]$$

```
> x:=backsub(augment(R,evalm(transpose(Q)&*b)));
```

$$x := \left[\frac{900991062}{756297791}, \frac{70597942}{756297791}, \frac{-543251988}{756297791}, \frac{-832910817}{756297791}, \frac{62544990}{756297791}\right]$$

```
> check:=evalm(A&*x-b);
```

$$check := [0, 0, 0, 0, 0]$$

In the next example we will use larger matrices and suppress the printout of the matrices. We only print the final results.

```
A:=randmatrix(15,15):   b:=randvector(15):
> R:=QRdecomp(A,Q='Q'):
> x:=backsub(augment(R,transpose(Q)&*b));
```

$$x := \left[\frac{6252799665699196799812589230 6431}{234243812997301475343932251020 64}, \frac{-1151433443111638244[illegible]828097979}{117121906498650737671966125510 32},\right.$$

$$\frac{-4158355555981406613227529482 1877}{234243812997301475343932251020 64}, \frac{29430272223266545757[illegible]8394141}{117121906498650737671966125510 32},$$

$$\frac{-4661300148597273060100628491857}{23424381299730147534393225102064}, \frac{-2401790680138335135718440601211}{11712190649865073767196612551032},$$

$$\frac{-6380148149456600852230373638 5701}{234243812997301475343932251020 64}, \frac{-4491725117759436457152532888761}{11712190649865073767196612551032},$$

$\frac{19796284681203190262853028511973}{23424381299730147534393225102064}$, $\frac{12115166588445338722314719104615}{5856095324932536883598306275516}$,
$\frac{7311986745858545501213998712017}{23424381299730147534393225102064}$, $\frac{-12916274918239871231210079994823}{23424381299730147534393225102064}$,
$\frac{-44956817717811196532816317428981}{23424381299730147534393225102064}$, $\frac{-14733962411569027091331464845787}{23424381299730147534393225102064}$,
$\frac{-17937643428226401930549382953961}{23424381299730147534393225102064}$]

```
> check:=evalm(A&*x-b);
```

$$check := [0, 0, 0, 0, 0, 0, 0, 0, 0, 0, 0, 0, 0, 0, 0]$$

7. Singular Value Decomposition.

The singular value decomposition provides a great deal of useful information about the matrix A. The maple command to perform the decomposition is **Svd.** Let A be an m by n matrix. Let U be an m by m orthonormal matrix and V an n by n orthonormal matrix. Let S be an m by n diagonal matrix with diagonal elements $s_1 \geq s_2 \geq \ldots \geq s_n \geq 0$. Then the singular value decomposition of A is $A = USV^T$. Furthermore

1). If $p \leq n$ of the singular values (the s's) are nonzero then the first p columns of U provide an orthonormal basis for the column space of A.

2). The first p columns of V form an orthonormal basis for the row space of A.

3). The last n-p columns of V form an orthonormal basis for the null space of A.

4). The last m-p columns of U form an orthonormal basis for the null space of A^T.

5). The Euclidean (2) norm of A is given by the largest value among the s's.

We can solve the matrix equation Ax = B using

$$USV^T x = b$$
$$Sy = U^T b.$$
$$x = Vy$$

Example 7.1

Generate a 5 by 5 random matrix A and vector b of length 5 and the use the singular value decomposition to solve the matrix equation Ax = b.

```
> A:=randmatrix(5,5);  b:=randvector(5);
```

$$A := \begin{bmatrix} -85 & -55 & -37 & -35 & 97 \\ 50 & 79 & 56 & 49 & 63 \\ 57 & -59 & 45 & -8 & -93 \\ 92 & 43 & -62 & 77 & 66 \\ 54 & -5 & 99 & -61 & -50 \end{bmatrix}$$

$$b := [-12, -18, 31, -26, -62]$$

```
> U:=matrix(5,5,0):   V:=copy(U):  q:=vector(5,0):
>s1:=evalf(Svd(A,U,V));
 S:=diag(s1[1],s1[2],s1[3],s1[4],s1[5]);
```

$$s1 := [215.3159815, 192.8527895, 110.7852504, 66.73636029, 33.05625793]$$

$$S := \begin{bmatrix} 215.3159815 & 0 & 0 & 0 & 0 \\ 0 & 192.8527895 & 0 & 0 & 0 \\ 0 & 0 & 110.7852504 & 0 & 0 \\ 0 & 0 & 0 & 66.73636029 & 0 \\ 0 & 0 & 0 & 0 & 33.05625793 \end{bmatrix}$$

```
> print(U,V);
```

$$\begin{bmatrix} -0.4003264629 & -0.5677384250 & 0.2827003489 & -0.6398772842 & -0.1674794847 \\ -0.1795173293 & 0.5587871806 & 0.6288251625 & 0.02763076550 & -0.5092601212 \\ 0.5539553017 & 0.08272047228 & -0.3796042972 & -0.4315116156 & -0.5966482635 \\ -0.4366877346 & 0.5809268674 & -0.3933775844 & -0.4890724096 & 0.2790878920 \\ 0.5567311718 & 0.1452971638 & 0.4751989913 & -0.4054620891 & 0.5279446462 \end{bmatrix}$$

$$\begin{bmatrix} 0.2160338793 , & 0.7373681815 , & -0.2234567203 , & -0.5351605781 , & 0.2707171543 \\ -0.2155367404 , & 0.4912693381 , & 0.3360924941 , & 0.6568003270 , & 0.4096993007 \\ 0.5195997159 , & 0.1783110248 , & 0.7140500468 , & -0.06013959605 , & -0.4298105087 \\ -0.3102515750 , & 0.4275691056 , & -0.3188371848 , & 0.2139211251 , & -0.7573039165 \\ -0.7352777516 , & 0.01823295529 , & 0.4749571663 , & -0.4825323485 , & -0.02474692053 \end{bmatrix}$$

```
> check:=evalm(A-U&*S&*transpose(V));
```

$$check := \begin{bmatrix} 0. & 0. & -0.1\ 10^{-7} & 0.2\ 10^{-7} & 0. \\ 0.2\ 10^{-7} & 0. & -0.2\ 10^{-7} & 0.4\ 10^{-7} & 0. \\ 0. & 0. & 0.3\ 10^{-7} & -0.2\ 10^{-7} & 0.2\ 10^{-7} \\ -0.1\ 10^{-7} & -0.1\ 10^{-7} & 0.1\ 10^{-7} & 0. & 0.2\ 10^{-7} \\ 0.2\ 10^{-7} & -0.8\ 10^{-8} & 0.1\ 10^{-7} & 0. & -0.1\ 10^{-7} \end{bmatrix}$$

```
> y:=backsub(augment(S,evalm(transpose(U)&*b)));
```

$$y := [0.009494846902, -0.1285617713, -0.4126317396, 0.4743867598, -1.431151509]$$

```
> x:=evalm(V&*y);
```

$$x := [-0.6418511722, -0.4786517685, 0.2739643556, 1.258945502, -0.3987981771]$$

```
> check:=evalm(A&*x-b);
```

$$check := [0., 0.3\ 10^{-7}, -0.3\ 10^{-7}, 0.2\ 10^{-7}, -0.2\ 10^{-7}]$$

Example 7.2

Same as in problem 7.1 but use a size of 10.

```
A:=randmatrix(10,10):   b:=randvector(10):
> U:=matrix(10,10,0):   V:=copy(U):  q:=vector(10,0):
```

```
> s1:=evalf(Svd(A,U,V)); S:=diag(1,1,1,1,1,1,1,1,1,1):
```

$s1 := [285.2009527, 247.6520534, 225.7830149, 203.5865991, 160.4006683,$
$151.2161910, 74.35166779, 60.97378580, 39.81865509, 8.197235759]$

```
> for j from 1 to 10 do S[j,j]:=s1[j]; od:
> print(U,V);
```

[0.6356072671 , -0.2938638343 , -0.1508664358 , 0.05153254196 , 0.2665335121 ,
0.1480306812 , -0.1981987341 , 0.2384899860 , 0.4217813275 , -0.3423715365]
[0.4956620992 , -0.1145676125 , -0.08202425898 , 0.2251002909 , 0.2279612936 ,
0.1720517044 , 0.1729236819 , -0.1701158886 , -0.6804283315 , 0.2835530912]
[0.01108709366 , 0.3170479805 , -0.5510846638 , -0.06645563068 ,
0.09623985985 , -0.03656715787 , 0.7231334979 , 0.06007503371 , 0.1042252225 ,
-0.2079753995]
[-0.3951320379 , -0.2086119784 , -0.2025685753 , 0.2891753557 , 0.7037671217 ,
-0.1015583095 , -0.08489347808 , -0.3074761194 , 0.2019533671 , 0.1660103012]
[0.3193884035 , 0.2961741358 , -0.06293222663 , -0.1843704104 , 0.1247028498 ,
-0.6839006412 , -0.1430233516 , 0.1489006881 , 0.1248732661 , 0.4804453156]
[0.1113292157 , -0.4924265828 , 0.2615481449 , 0.3534062586 , -0.3046334603 ,
-0.4126687378 , 0.4573305131 , -0.1800263380 , 0.2154792485 , 0.02700798170]
[0.05808607675 , 0.1861343241 , 0.3211711768 , 0.08781776983 , 0.06842505234
, 0.5116578706 , 0.2744575040 , 0.1567323039 , 0.4166341282 , 0.5578220486]
[0.1443155947 , -0.2095671226 , -0.03929723992 , -0.6977115649 ,
-0.02435845151 , 0.1187820889 , 0.06591402640 , -0.6260499841 , 0.1549592574 ,
0.1091366007]
[0.2312296607 , 0.5554129528 , 0.04643362992 , 0.4088130778 , -0.1451454300 ,
0.02211125960 , -0.1846214856 , -0.5873176760 , 0.1844136977 , -0.1848605428]
[0.04470960235 , 0.1980630144 , 0.6682543139 , -0.1952064701 , 0.4875945034 ,
-0.1497545928 , 0.2394204524 , 0.002447418607 , -0.1176448838 , -0.3778351498
],
[-0.04199305163 , -0.6387287778 , -0.009675347005 , 0.2616193914 ,
0.2694613401 , 0.3738941830 , -0.4578199213 , 0.09757687219 , -0.2009414250 ,
0.2232033987]
[-0.3091156119 , -0.4916523355 , 0.2047014934 , -0.2601237459 , -0.07324255945
, -0.1563622051 , -0.09520210227 , -0.2657413921 , 0.5312061064 , -0.4018494592
]
[-0.5439344876 , -0.03591020997 , -0.1213813867 , 0.5521320902 ,
-0.06570371698 , -0.009198708324 , 0.4127559496 , 0.01000851056 ,
0.2969123051 , 0.3467510464]
[-0.3505757408 , 0.3239529289 , -0.1393257984 , 0.07633636426 , 0.3425138969 ,
-0.1747309113 , -0.4116499841 , 0.5611111840 , 0.1663811386 , -0.2950956063]
[-0.07847921942 , -0.2719099478 , 0.5252521589 , -0.2360018636 ,

0.07257853708 , -0.2657319021 , 0.3592911560 , 0.5562619305 , -0.2200102999 , 0.1597378866]
[-0.07081960223 , 0.1438918456 , 0.3629843345 , 0.1444108134 , -0.7152616794 , 0.4145630437 , -0.2237447426 , 0.2690427128 , 0.05553478007 , -0.1125878338]
[0.3168845044 , -0.2919442042 , -0.3697501825 , 0.1935283898 , -0.4446094731 , -0.5877124610 , -0.1824750766 , 0.2289950337 , 0.06367956765 , 0.08549778147]
[-0.4004159900 , -0.1573490282 , -0.3050159967 , 0.01705494179 , -0.1972528123 , -0.007469453284 , 0.1844258345 , -0.04921075006 , -0.6408334395 , -0.4853000523]
[-0.2370714838 , 0.1912937795 , 0.4473005545 , 0.2190086284 , -0.02035163804 , -0.4640187097 , -0.3873587884 , -0.4101176662 , -0.3071859609 , 0.1755765884]
[0.3947024180 , -0.05334002335 , 0.2939104761 , 0.6252628204 , 0.2156988802 , -0.04344467264 , 0.2121451825 , -0.006789347792 , 0.03207137437 , -0.5191668204]

```
> check:=evalm(A-U&*S&*transpose(V));
```

check :=

[-0.1 10^{-7} , -0.2 10^{-7} , -0.1 10^{-7} , 0. , -0.2 10^{-7} , 0. , 0.2 10^{-7} , -0.2 10^{-7} , -0.1 10^{-7} , -0.1 10^{-7}]
[0.2 10^{-7} , -0.2 10^{-7} , -0.1 10^{-7} , -0.1 10^{-7} , 0. , 0. , 0.1 10^{-7} , -0.2 10^{-7} , -0.1 10^{-7} , 0.1 10^{-7}]
[0. , 0. , -0.1 10^{-7} , -0.1 10^{-7} , 0.3 10^{-7} , 0.1 10^{-7} , -0.13 10^{-7} , -0.3 10^{-7} , 0.2 10^{-7} , 0.2 10^{-7}]
[0.2 10^{-7} , 0. , -0.3 10^{-7} , 0. , 0.2 10^{-7} , -0.6 10^{-7} , -0.2 10^{-7} , -0.4 10^{-7} , 0. , 0.45 10^{-7}]
[-0.3 10^{-7} , -0.2 10^{-7} , 0. , 0. , 0.3 10^{-8} , 0.1 10^{-7} , 0.3 10^{-7} , 0. , -0.1 10^{-7} , -0.20 10^{-7}]
[-0.2 10^{-7} , 0. , 0.1 10^{-7} , 0. , 0. , 0. , 0.2 10^{-7} , -0.15 10^{-7} , 0. , 0.]
[0.11 10^{-7} , -0.2 10^{-7} , 0.2 10^{-8} , 0. , -0.2 10^{-7} , 0.1 10^{-7} , -0.1 10^{-7} , -0.1 10^{-7} , 0. , -0.1 10^{-7}]
[0.7 10^{-8} , 0.1 10^{-7} , 0.1 10^{-7} , 0. , -0.1 10^{-7} , -0.1 10^{-7} , -0.1 10^{-7} , 0.5 10^{-8} , 0. , 0.1 10^{-7}]
[-0.1 10^{-7} , -0.4 10^{-7} , 0.101 10^{-7} , 0.14 10^{-7} , -0.2 10^{-7} , 0.4 10^{-7} , 0.10 10^{-7} , 0.1 10^{-7} , 0. , 0.1 10^{-7}]
[-0.1 10^{-7} , 0.19 10^{-7} , -0.1 10^{-7} , 0.1 10^{-7} , 0. , -0.1 10^{-7} , 0.2 10^{-7} , 0.1 10^{-7} , 0. , -0.3 10^{-7}]

```
> y:=backsub(augment(S,evalm(transpose(U)&*b)));
```

y := [0.1384144392, 0.2699838652, -0.3525829225, 0.4048539428, -0.06942288426, 0.5275589926, -0.08385650941, 0.7312187898, 0.4319479897, -9.383597959]

```
> x:=evalm(V&*y);
```

$x := [-1.961891427, 3.383499177, -2.971760826, 3.288748227, -1.727571888,$
$1.523849865, -0.6974507762, 4.251901153, -2.341227949, 5.014601434]$

```
> check:=evalm(A&*x-b);
```

check :=

$[0.2\ 10^{-6}, 0.1\ 10^{-6}, 0.2\ 10^{-6}, -0.1\ 10^{-7}, 0.11\ 10^{-6}, 0., 0.1\ 10^{-6}, 0., 0.2\ 10^{-6}, -0.1\ 10^{-6}]$

Exercises for chapter 4

1). Generate random symmetric matrices of size 3 by 3, 5 by 5, 7 by 7 and 9 by 9. For
each matrix answer the following questions.
a). find all of the eigenvalues and eigenvectors in floating point form.
b). Form the matrix P of eigenvectors and the diagonal matrix D containing the eigenvalues.
c). determine how close to the zero matrix AP – PD is.
d). determine how close to the zero matrix $A - PDP^T$ is.

2). For each of the matrices listed below answer the following questions.

$$\begin{bmatrix} 19 & 17 & 12 \\ -36 & -34 & -26 \\ 21 & 21 & 18 \end{bmatrix}$$

$$\begin{bmatrix} -48 & 25 & 8 & 9 & 12 \\ -73 & 38 & 14 & 15 & 18 \\ 50 & -26 & -12 & -12 & -12 \\ 1 & -1 & 4 & 4 & 0 \\ -62 & 32 & 4 & 6 & 16 \end{bmatrix}$$

$$\begin{bmatrix} -24 & -17 & 3 & 4 & 5 & 6 & 7 \\ 44 & 31 & -4 & -6 & -8 & -10 & -12 \\ 10 & 6 & 1 & -2 & -2 & -2 & -2 \\ 32 & 20 & -2 & 0 & -6 & -8 & -10 \\ -43 & -25 & 5 & 6 & 12 & 8 & 9 \\ 16 & 9 & -2 & -2 & -2 & 4 & -2 \\ -5 & -3 & 1 & 0 & -1 & -2 & 4 \end{bmatrix}$$

$$\begin{bmatrix} 21 & 16 & 3 & 4 & 5 & 7 & 8 & 9 & 10 \\ -44 & -31 & -4 & -6 & -8 & -12 & -14 & -16 & -18 \\ 4 & 4 & -1 & 2 & 2 & 2 & 2 & 2 & 2 \\ 42 & 26 & 2 & 2 & 6 & 10 & 12 & 14 & 16 \\ -66 & -44 & -5 & -8 & -12 & -17 & -20 & -23 & -26 \\ 68 & 48 & 5 & 8 & 11 & 18 & 20 & 23 & 26 \\ -13 & -15 & -5 & -6 & -7 & -9 & -8 & -11 & -12 \\ 34 & 20 & -1 & 0 & 1 & 3 & 4 & 8 & 6 \\ -37 & -22 & 1 & 0 & -1 & -3 & -4 & -5 & -2 \end{bmatrix}$$

2). For each matrix answer the following questions.

a). find all of the eigenvalues and eigenvectors in rational form.

b). Form the matrix P of eigenvectors and the diagonal matrix D containing the eigenvalues.

c). determine how close to the zero matrix AP – PD is.

d). determine how close to the zero matrix $A - PDP^T$ is.

Project 1.

Given the matrix

$$\begin{bmatrix} -17 & -14 & 7 & 9 & 12 \\ 42 & 37 & -15 & -21 & -30 \\ -42 & -38 & 14 & 21 & 30 \\ 36 & 28 & -14 & -17 & -24 \\ 21 & 24 & -9 & -15 & -20 \end{bmatrix}$$

answer the following questions:

1). Find the characteristic polynomial.

2). Find the roots of the polynomial in (1).

3). Using the roots in (2) find the corresponding eigenvectors.

4). Use the eigenvalues to form a diagonal matrix D and use the eigenvectors to form the matrix of eigenvectors P.

5). Arrange the eigenvalues in D and the eigenvectors in P so that AP = PD.

Project 2.

Given the matrix

$$\begin{bmatrix} -17 & -14 & 7 & 9 & 12 \\ 42 & 37 & -15 & -21 & -30 \\ -42 & -38 & 14 & 21 & 30 \\ 36 & 28 & -14 & -17 & -24 \\ 21 & 24 & -9 & -15 & -20 \end{bmatrix}$$

do the following:

1). Modify the Gram-Schmidt procedure in the text so that it will produce rational and radical entries rather than floating point numbers.

2). Use the modified procedure to find the matrix Q containing the orthonormal vectors which span the same space as the columns of the above matrix.

3). Show that the span of the columns of the above matrix is the same as the span of the columns of the matrix Q.

4). Given the vector v = [2,3,0,-5,1] and find the expansion of v as a linear combination of the given matrix A. Also find the expansion of v as a linear combination of the columns of Q.

Project 3.

Given the matrix

$$S=\begin{bmatrix} 1 & 1 & -2 & 2 & 2 & -2 & 2 \\ 1 & 2 & 0 & 1 & 2 & -3 & 0 \\ -2 & 0 & 9 & -5 & -6 & 3 & -9 \\ 2 & 1 & -5 & 10 & -2 & 6 & 13 \\ 2 & 2 & -6 & -2 & 12 & -14 & -2 \\ -2 & -3 & 3 & 6 & -14 & 22 & 13 \\ 2 & 0 & -9 & 13 & -2 & 13 & 25 \end{bmatrix}$$

do the following:

(1). Find a basis for the column space of S.

(2). Using maple functions find the S = QR decomposition and use it to find an orthonormal basis for the column space of S.

(3). Show that this orthonormal basis spans the column space of S.

(4). Find a basis for the null space of S.

(5). Using the QR decomposition find an orthonormal basis for the null space of S.

(6). Show that the orthonormal basis in (5) actually spans the null space of S.

Project 4.

Given the matrix

$$A = \begin{bmatrix} \frac{43}{9} & \frac{-10}{9} & 0 & \frac{2}{3} & \frac{2}{9} \\ \frac{-10}{9} & \frac{397}{90} & \frac{2}{5} & \frac{3}{10} & \frac{-43}{45} \\ 0 & \frac{2}{5} & \frac{16}{5} & \frac{2}{5} & \frac{4}{5} \\ \frac{2}{3} & \frac{3}{10} & \frac{2}{5} & \frac{119}{30} & \frac{-11}{15} \\ \frac{2}{9} & \frac{-43}{45} & \frac{4}{5} & \frac{-11}{15} & \frac{164}{45} \end{bmatrix}$$

answer the following questions.

(1). Is A a positive definite matrix? Justify your answer.

(2). Find the eigenvalues and eigenvectors of A.

(3). Normalize the eigenvectors so that they have unit length.

(4). Any vector $u \in R^5$ can be written as a linear combination of the unit eigenvectors in (3) as follows $u = \sum_{j=1}^{5} c_j v_j$. Show that $c_m = v_m^T u$.

(5). Using the general expansion $u = \sum_{j=1}^{5} c_j v_j$. determine the form of $u^T Au$.

(6). Using (5) determine $u^T Au$ where u = [1,0,3,-1,2].

Project 5.

Given the matrix

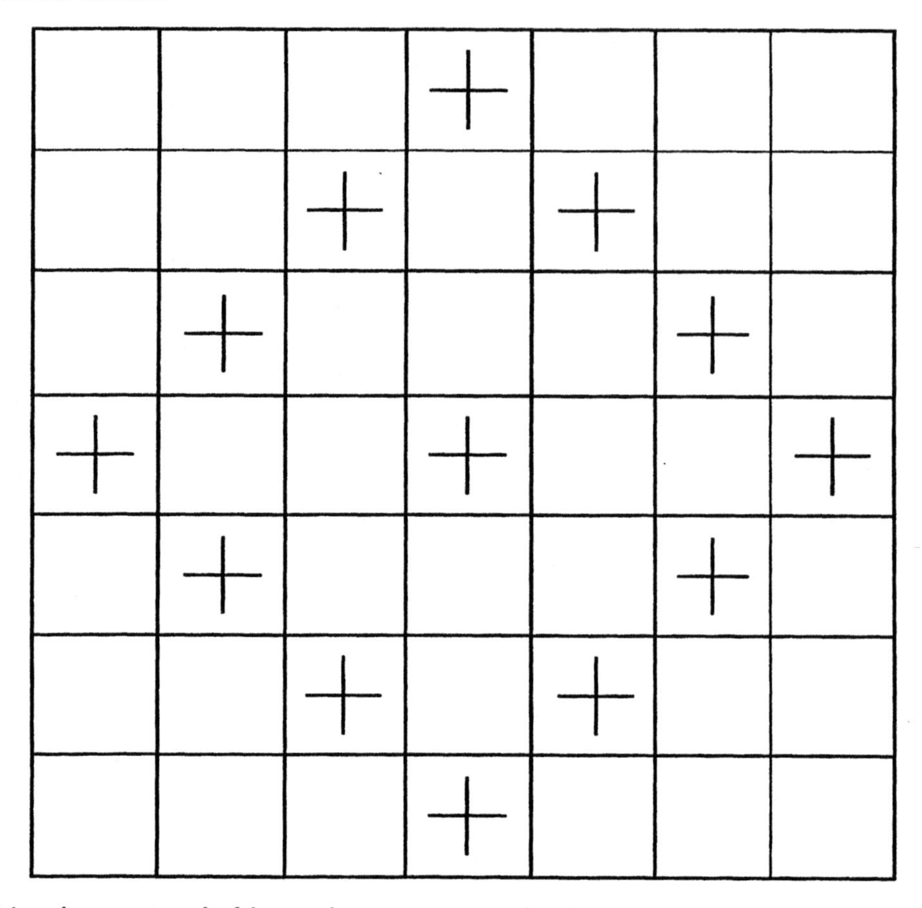

our object is to transmit this matrix to some one else by electronic means using the smallest amount of information that we can get by with. We should note that we are dealing with a symmetric matrix.

The first thing that we shall do is to encode our drawing in a 7 by 7 symmetric matrix by entering a one if the cell contains a cross and zero otherwise. Our matrix is then

$$A = \begin{bmatrix} 0 & 0 & 0 & 1 & 0 & 0 & 0 \\ 0 & 0 & 1 & 0 & 1 & 0 & 0 \\ 0 & 1 & 0 & 0 & 0 & 1 & 0 \\ 1 & 1 & 1 & 1 & 1 & 1 & 1 \\ 0 & 1 & 0 & 0 & 0 & 1 & 0 \\ 0 & 0 & 1 & 0 & 1 & 0 & 0 \\ 0 & 0 & 0 & 1 & 0 & 0 & 0 \end{bmatrix}$$

Now let (λ_j, p_j) $j = 1,2,\ldots,7$ be the eigenvalues and eigenvectors of the matrix A. The spectral decomposition of A is

$A = \lambda_1 p_1 p_1^T + \lambda_2 p_2 p_2^T + \cdots + \lambda_7 p_7 p_7^T$. One way of transmitting our picture would be to transmit all 7 eigenvalues and all 7 eigenvectors then the recipient could reconstruct the spectral decomposition and hence A. Our question is, is it possible to transmit the picture accurately by sending less than all of the eigenvalues and eigenvectors? We shall assume that $|\lambda_1| \ge |\lambda_2| \ge \cdots \ge |\lambda_7|$.

To answer this question we need to do the following:

(1). Find the 7 eigenvalues and eigenvectors as floating point numbers unless they are all rational.

(2). Create the matrices $B_j = \lambda_j p_j p_j^T$ $j = 1,2,\ldots,7$.

(3). Create the matrices $C_m = \sum_{j=1}^{m} B_j$ $m = 1,2,\ldots,7$.

(4). For each of the matrices C_m create the matrix $F[i,j] = \begin{cases} 0 & \text{if } C_m[i,j] < 0.5 \\ 1 & \text{if } C_m \ge 0.5 \end{cases}$

and determine the smallest value of m the gives F = A.

(5). Can the drawing be successfully transmitted by sending less than the full number of eigenvalues and eigenvectors.

Chapter 4.1
Eigenvalues, Eigenvectors, and Orthogonal Vectors

Let A be an n by n matrix and x a vector and let us ask the question when will Ax be parallel to x. That is when will Ax be in either the same or exactly opposite direction to x? That question leads us to the following definition.

Definition 4.0
Let A be an n by n matrix, x a nonzero vector and λ a scalar then λ is an eigenvalue of A and x its corresponding non zero eigenvector if $Ax = \lambda x$.

Thus if Ax is parallel to the vector x then x is an eigenvector of A. Eigenvalues can be complex valued. Some properties of eigenvalues that we will be using in this chapter are:

Properties
1). If A is a symmetric matrix then all of its eigenvalues are real valued.

2). If A is a symmetric matrix then eigenvectors corresponding to distinct eigenvalues are orthogonal.

3). Eigenvectors are not unique for if $Ax = \lambda x$ and c is any nonzero scalar then $A(cx) = \lambda(cx)$. Even if we require x to be of unit length they are still not unique since there will be two of them.

Now we turn our attention to finding the eigenvalues of a matrix A.

Finding Eigenvalues and Eigenvectors
We wish to find the eigenvalues and eigenvectors of an n by n matrix A. That is all the solutions to the matrix equation $Ax = \lambda x$ or $Ax = \lambda Ix$ where I is the n by n identity matrix. This matrix equation can be written as $Ax - \lambda Ix = 0$ or $(A - \lambda I)x = 0$. This last equation is a homogeneous matrix equation. We know from the theory of linear algebra that a homogeneous matrix equation has a non zero solution if and only if the determinant of the underlying matrix is zero. This means that the eigenvalues A are the solutions of the characteristic equation that is $\det(A - \lambda I) = 0$.

The characteristic equation is a polynomial of degree n in λ. As such it has n roots some of which may be real, complex, or repeated. If p(x) is a polynomial of degree n and a particular root, say r, appears m times in the list of all n roots then r is said to have multiplicity m. We can find the characteristic polynomial of a matrix A by using the maple function CharacteristicPolynomial.

```
> restart; with(LinearAlgebra):
```

```
> A:=RandomMatrix(5,5,outputoptions=[shape=symmetric]);
```

$$A := \begin{bmatrix} 28 & -71 & -79 & 62 & 30 \\ -71 & -50 & -8 & -56 & -21 \\ -79 & -8 & -90 & -62 & -34 \\ 62 & -56 & -62 & 16 & -7 \\ 30 & -21 & -34 & -7 & 20 \end{bmatrix}$$

```
> p:=CharacteristicPolynomial(A,lambda);
```

$$p := \lambda^5 + 76\,\lambda^4 - 27848\,\lambda^3 - 2333782\,\lambda^2 - 3780865\,\lambda + 1987651984$$

How can we find the roots of this polynomial? We can use maple's solve **or** fsolve **command. Using these commands we have:**

```
> w:=fsolve(p=0,lambda);
```

$$w := -162.7126172, -58.38129639, -49.08359542, 25.23272984, 168.9447792$$

To find the individual eigenvectors we proceed as follows.

```
> I1:=DiagonalMatrix(<1,1,1,1,1>):
zero:=Vector([0,0,0,0,0]):
> x:=LinearSolve(A-w[1]*I1,zero);
```

$$x := \begin{bmatrix} 0. \\ 0. \\ 0. \\ 0. \\ 0. \end{bmatrix}$$

We use I1 instead of I because maple uses I in complex numbers. W[1] = -162.7126172 is the first of the eigenvalues listed in w. The solution for x is zero indicating that there are no non zero solutions. This does not mean that w[1] fails to be an eigenvalue, it simply indicates the difficulty of computing with floating point numbers. W[1] is a ten digit approximation **to an eignevalue which may take an infinite number of digits to specify accurately.**

How can we get around this problem?

2. The Maple commands Eigenvals and Eigenvectors

The maple command Eigenvectors **will compute numerical approximations to the eigenvalues and eigenvectors provided that at least one of the entries in the matrix A is in floating point. To satisfy this condition we will change the entry A[1,1] to floating point.**

```
>A[1,1]:=-85.0;
```

$$A_{1,1} := -85.0$$

Now that we have met that requirement we should be able to get all of the information that we want from:

```
>w1:=Eigenvectors(A,output='list');
```

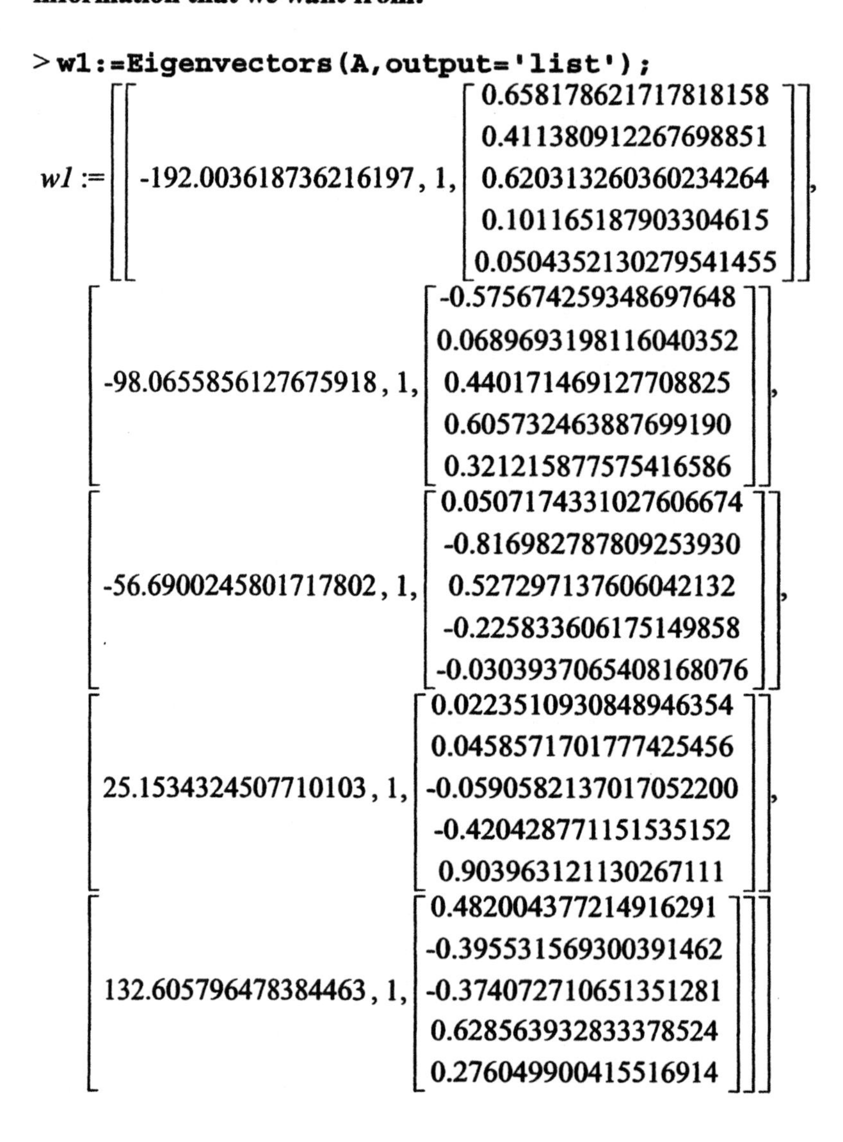

$$w1 := \left[\left[-192.003618736216197, 1, \left\{\begin{bmatrix} 0.658178621717818158 \\ 0.411380912267698851 \\ 0.620313260360234264 \\ 0.101165187903304615 \\ 0.0504352130279541455 \end{bmatrix}\right\}\right], \left[-98.0655856127675918, 1, \left\{\begin{bmatrix} -0.575674259348697648 \\ 0.0689693198116040352 \\ 0.440171469127708825 \\ 0.605732463887699190 \\ 0.321215877575416586 \end{bmatrix}\right\}\right], \left[-56.6900245801717802, 1, \left\{\begin{bmatrix} 0.0507174331027606674 \\ -0.816982787809253930 \\ 0.527297137606042132 \\ -0.225833606175149858 \\ -0.0303937065408168076 \end{bmatrix}\right\}\right], \left[25.1534324507710103, 1, \left\{\begin{bmatrix} 0.0223510930848946354 \\ 0.0458571701777425456 \\ -0.0590582137017052200 \\ -0.420428771151535152 \\ 0.903963121130267111 \end{bmatrix}\right\}\right], \left[132.605796478384463, 1, \left\{\begin{bmatrix} 0.482004377214916291 \\ -0.395531569300391462 \\ -0.374072710651351281 \\ 0.628563932833378524 \\ 0.276049900415516914 \end{bmatrix}\right\}\right]\right]$$

Starting at the topmost left we have listed the eigenvalue –192.003618736216197, next the number one indicates that the multiplicity of this eigenvalue is one. This information is followed by a set giving all of the associated eigenvectors, in this case one. If the multiplicity had been 3 then the set would have contained 3 eigenvectors (since A is symmetric). Thus

$\lambda = -192.003618736216197$ and

$$x\_1 := \begin{bmatrix} 0.6581786217178158 \\ 0.411380912267698851 \\ 0.620313260360234264 \\ 0.101165187903304615 \\ 0.0504352130279541455 \end{bmatrix}$$

These numbers are long and tedious to type. There is a maple function op **that can help us. To get the pieces that we want. we may have to experiment around as in:**

```
> nops(w1[1]); # how many operands in w1?
```

3

```
> lambda_1:=op(1,w1[1]);
```

$lambda\_1 := -192.003618736216197$

```
> m_1:=op(2,w1[1]);
```

$m\_1 := 1$

```
> x_1:=op(3,w1[1]);
```

$$x\_1 := \begin{bmatrix} 0.6581786217178158 \\ 0.411380912267698851 \\ 0.620313260360234264 \\ 0.101165187903304615 \\ 0.0504352130279541455 \end{bmatrix}$$

The first line tells us that there are 3 operands in the first component of w1. The second line accesses the eigenvalue while the line after that tells us that the multiplicity is 1. We accessed this for illustrative purposes only since we would normally not need it. Finally the solution for x_1 is the associated eigenvector. Finally we use these commands to pick out the eigenvalue and eigenvector in w1[5] .

```
> x_1:=op(1,w1[5]);
```

$x\_1 := 132.605796478384463$

```
> lambda_1:=op(2,w1[5]);
```

$lambda\_1 := 1$

```
> x_1:=op(3,w1[5]);
```

$$x\_1 := \begin{bmatrix} 0.482004377214916291 \\ -0.395531569300391462 \\ -0.374072710651351281 \\ 0.628563932833378524 \\ 0.276049900415516914 \end{bmatrix}$$

Now lets look at a case where the eigenvalues are integers so that we can solve exactly for the eigenvectors.

```
> A:=Matrix([[-9,21,7,9,13],[-4,5,4,4,4],[18,-39,-8,-14,-
22],[-26,42,14,22,26],[8,6,-8,-8,0]]);
```

$$A := \begin{bmatrix} -9 & 21 & 7 & 9 & 13 \\ -4 & 5 & 4 & 4 & 4 \\ 18 & -39 & -8 & -14 & -22 \\ -26 & 42 & 14 & 22 & 26 \\ 8 & 6 & -8 & -8 & 0 \end{bmatrix}$$

For illustrative purposes only we will show how to solve for the eigenvalues and eigenvectors without using the maple functions Eigenvals **and** Eigenvectors. **These two commands are by far the most efficient way of coming up with the desired information. The more primitive methods that we are going to employ are for the illustration of certain maple commands only.**

```
> p:=CharacteristicPolynomial(A,lambda);
```

$$p := \lambda^5 - 10\,\lambda^4 + 3\,\lambda^3 + 118\,\lambda^2 - 88\,\lambda - 192$$

```
> w:=solve(p=0,lambda);
```

$$w := -1, 2, -3, 4, 8$$

```
> I1:=DiagonalMatrix(<1,1,1,1,1>):
zero:=Vector([0,0,0,0,0]):
```

Again we use I1 rather than I since I is a reserved word in maple. Now we can solve for the eigenvector associated with the eigenvalue w[1] = -1.

```
> x:=LinearSolve(A-w[1]*I1,zero);
```

$$x := \begin{bmatrix} \frac{1}{2}\_t0_4 \\ \_t0_4 \\ 0 \\ \_t0_4 \\ -2\,\_t0_4 \end{bmatrix}$$

```
> x1:=map2(subs,_t0[4]=1,x); # x1 is an eigenvector for
w[1].
```

$$x1 := \begin{bmatrix} \frac{1}{2} \\ 1 \\ 0 \\ 1 \\ -2 \end{bmatrix}$$

In this way we can find all of the eigenvectors. We will limit ourselves to the additional eigenvalue
W[5] = 8.

```
> x:=LinearSolve(A-w[5]*I1,zero);
```

$$x := \begin{bmatrix} \_t0_5 \\ 0 \\ -2\_t0_5 \\ 2\_t0_5 \\ \_t0_5 \end{bmatrix}$$

```
> x5:=map2(subs,_t0[5]=1,x); # x5 is an eigenvector for
w[5].
```

$$x5 := \begin{bmatrix} 1 \\ 0 \\ -2 \\ 2 \\ 1 \end{bmatrix}$$

Notice that we can solve for the eigenvectors using this method only because we have the exact eigenvalue and not just a 10 digit approximation.

3. Diagonalization and powers

Suppose that $Ax_i = \lambda_i x_i \ i = 1,2,3,\ldots,n$ **and let** $P = [x_1, x_2, \ldots, x_n]$ **be the matrix whose columns are the eigenvectors** $x_i \ i = 1,2,\ldots,n$**. Furthermore let** $D = diag(\lambda_1, \lambda_2, \ldots, \lambda_n)$ **be a diagonal matrix with the eigenvalues of the matrix A on the main diagonal. Then we have**
AP = PD or $A = PDP^{-1}$**. If this can be done then we say that the matrix A can be diagonalized or that A is similar to a diagonal matrix D.**

Not every matrix can be diagonalized. Every symmetric matrix can be diagonalized. **What happens in the case where a matrix cannot be diagonalized? It can be shown that every matrix A is similar to its Jordan canonical form. The Jordan canonical form looks like**

$$J = \begin{bmatrix} J1 & 0 & 0 \\ . & . & . \\ 0 & 0 & Jl \end{bmatrix}$$

in block form where each block Jk looks like

$$Jk = \begin{bmatrix} \lambda & 1 & 0 & 0 & 0 \\ 0 & \lambda & 1 & 0 & 0 \\ . & . & . & . & . \\ 0 & 0 & 0 & \lambda & 1 \\ 0 & 0 & 0 & 0 & \lambda \end{bmatrix}$$

where we have ones on the super diagonal and the associated eigenvalue on the main diagonal. The sizes of the blocks vary and can be 1 by 1 in many cases. There is a maple function JordanForm **which will give us the decomposition. Consider the following example.**

```
> A:=Matrix([[-15,-11,1,1,5,6,6,7],[24,16,0,-2,-10,-12,-12,-
14],[34,24,-1,-2,-15,-18,-18,-21],[11,7,1,-3,-5,-6,-6,-
7],[-42,-24,1,3,18,18,18,21],[29,13,1,-3,-15,-14,-18,-
21],[-24,-13,0,2,10,12,16,15],[24,12,0,-2,-10,-12,-12,-
10]]);
```

$$A := \begin{bmatrix} -15 & -11 & 1 & 1 & 5 & 6 & 6 & 7 \\ 24 & 16 & 0 & -2 & -10 & -12 & -12 & -14 \\ 34 & 24 & -1 & -2 & -15 & -18 & -18 & -21 \\ 11 & 7 & 1 & -3 & -5 & -6 & -6 & -7 \\ -42 & -24 & 1 & 3 & 18 & 18 & 18 & 21 \\ 29 & 13 & 1 & -3 & -15 & -14 & -18 & -21 \\ -24 & -13 & 0 & 2 & 10 & 12 & 16 & 15 \\ 24 & 12 & 0 & -2 & -10 & -12 & -12 & -10 \end{bmatrix}$$

Now to find its Jordan decomposition we use.

```
> Q:=JordanForm(A,output='Q'):   # Here we have AQ = QJ.
```

```
> J:=Q^(-1).A.Q;
```

$$J := \begin{bmatrix} 3 & 0 & 0 & 0 & 0 & 0 & 0 & 0 \\ 0 & 4 & 1 & 0 & 0 & 0 & 0 & 0 \\ 0 & 0 & 4 & 0 & 0 & 0 & 0 & 0 \\ 0 & 0 & 0 & -2 & 1 & 0 & 0 & 0 \\ 0 & 0 & 0 & 0 & -2 & 1 & 0 & 0 \\ 0 & 0 & 0 & 0 & 0 & -2 & 0 & 0 \\ 0 & 0 & 0 & 0 & 0 & 0 & -2 & 0 \\ 0 & 0 & 0 & 0 & 0 & 0 & 0 & 4 \end{bmatrix}$$

Just for a quick check on our work we have.

```
> evalm(A&*Q-Q&*J);
```

$$\begin{bmatrix} 0 & 0 & 0 & 0 & 0 & 0 & 0 & 0 \\ 0 & 0 & 0 & 0 & 0 & 0 & 0 & 0 \\ 0 & 0 & 0 & 0 & 0 & 0 & 0 & 0 \\ 0 & 0 & 0 & 0 & 0 & 0 & 0 & 0 \\ 0 & 0 & 0 & 0 & 0 & 0 & 0 & 0 \\ 0 & 0 & 0 & 0 & 0 & 0 & 0 & 0 \\ 0 & 0 & 0 & 0 & 0 & 0 & 0 & 0 \\ 0 & 0 & 0 & 0 & 0 & 0 & 0 & 0 \end{bmatrix}$$

which shows that AP = PJ or alternatively $A = PJP^{-1}$.

It should be noted that the Jordan canonical form can only be computed accurately when all of the eigenvalues are known exactly. It can be shown that if we only have approximations to the eigenvalues then we cannot compute the Jordan canonical form with enough accuracy to be useful.

If A can be diagonalized then we can make use of this in computing powers of A. To see this note that

$$A = PDP^{-1} \Rightarrow$$
$$A^2 = \left(PDP^{-1}\right)\left(PDP^{-1}\right) = PD^2P^{-1}.$$

In general we can use mathematical induction to show that $A^n = PD^nP^{-1}$. As an example we have:

```
> restart; with(LinearAlgebra):
```

```
> A:=Matrix([[-9,21,7,9,13],[-4,5,4,4,4],[18,-39,-8,-14,-
22],[-26,42,14,22,26],[8,6,-8,-8,0]]);
```

$$A := \begin{bmatrix} -9 & 21 & 7 & 9 & 13 \\ -4 & 5 & 4 & 4 & 4 \\ 18 & -39 & -8 & -14 & -22 \\ -26 & 42 & 14 & 22 & 26 \\ 8 & 6 & -8 & -8 & 0 \end{bmatrix}$$

```
> w:=Eigenvectors(A,output='list');
```

$$w := \left[\left[2, 1, \left\{ \begin{bmatrix} -1 \\ 0 \\ 1 \\ -2 \\ 0 \end{bmatrix} \right\} \right], \left[-1, 1, \left\{ \begin{bmatrix} \frac{1}{2} \\ 1 \\ 0 \\ 1 \\ -2 \end{bmatrix} \right\} \right], \left[4, 1, \left\{ \begin{bmatrix} \frac{-1}{2} \\ 0 \\ 1 \\ \frac{-3}{2} \\ 0 \end{bmatrix} \right\} \right], \left[8, 1, \left\{ \begin{bmatrix} 1 \\ 0 \\ -2 \\ 2 \\ 1 \end{bmatrix} \right\} \right], \left[-3, 1, \left\{ \begin{bmatrix} 1 \\ 1 \\ -1 \\ 2 \\ -2 \end{bmatrix} \right\} \right] \right]$$

```
> D1:=DiagonalMatrix(<2,-1,4,8,-3>):
>
P:=<op(3,w[1])[1]|op(3,w[2])[1]|op(3,w[3])[1]|op(3,w[4])[1]
|op(3,w[5])[1]>;
```

$$P := \begin{bmatrix} -1 & \frac{1}{2} & \frac{-1}{2} & 1 & 1 \\ 0 & 1 & 0 & 0 & 1 \\ 1 & 0 & 1 & -2 & -1 \\ -2 & 1 & \frac{-3}{2} & 2 & 2 \\ 0 & -2 & 0 & 1 & -2 \end{bmatrix}$$

Now that we have AP = PD1 we shall use our method to compute $A^{20} = PD1^{20}P^{-1}$.

```
> A_20:=evalm(P.D1^(20).P^(-1));
```

$A\_20 :=$

[-2192047589599 , 2305842998752292175 , -6972520225 , 1092538058975 , 1152921497633278175]
[6973568800 , -10460353199 , -6973568800 , -6973568800 , -6973568800]
[4391070845150 , -4611686007965986125 , 6972520226 , -2192049686750 , -2305843002240125150]
[-6583118434750 , 4611685997504584350 , -13945040450 , 3284587745726 , 2305842995266556350]
[-13947137600 , 2305843030134400350 , 13947137600 , 13947137600 , 1152921518553984576]

Exactly the same technique can be used when the diagonal matrix D is replaced by the Jordan canonical form J. In this case computing J^n is not as efficient as computing D^n because J is upper triangular and not diagonal. Hence rather quickly the powers will fill up the zero positions in the upper triangular portion of J^n.

4. Positive Definite Matrices

Definition 4.1

The n by n symmetric matrix A is said to be positive definite if $x^T Ax > 0\ \forall x \neq 0$.

We shall need the following property of symmetric matrices.

Property

If A is an n by n symmetric matrix then there exists an orthonormal matrix P such that
$P^T P = I$ and a diagonal matrix D such that AP = PD. Thus D contains the eigenvalues and P the eigenvectors of A.

Theorem 4.1

The n by n symmetric matrix A is positive definite if and only if all of the eigenvalues of A are positive.

This is an easy test for us to apply using maple. To see this let us generate 3 random 5 by 5 matrices and check to see if they are positive definite or not.

```
> restart; with(LinearAlgebra):
> A:=RandomMatrix(5,5,outputoptions=[shape=symmetric]);
```

$$A := \begin{bmatrix} 28 & -71 & -79 & 62 & 30 \\ -71 & -50 & -8 & -56 & -21 \\ -79 & -8 & -90 & -62 & -34 \\ 62 & -56 & -62 & 16 & -7 \\ 30 & -21 & -34 & -7 & 20 \end{bmatrix}$$

```
> A[1,1]:=-85.0;   w:=Eigenvalues(A); # Not positive definite
```

$$A_{1,1} := -85.0$$

$$w := \begin{bmatrix} -192.003618736216225 \\ -98.0655856127675349 \\ -56.6900245801717518 \\ 25.1534324507710174 \\ 132.605796478384463 \end{bmatrix}$$

```
> A:=RandomMatrix(5,5,outputoptions=[shape=symmetric]);
```

$$A := \begin{bmatrix} -41 & -36 & 66 & 5 & -65 \\ -36 & 13 & 26 & 68 & 55 \\ 66 & 26 & -66 & -82 & 97 \\ 5 & 68 & -82 & 38 & -75 \\ -65 & 55 & 97 & -75 & 5 \end{bmatrix}$$

```
> A[1,1]:=92.0;   w:=Eigenvalues(A); # Not positive definite
```

$$A_{1,1} := 92.0$$

$$w := \begin{bmatrix} -171.090588778361763 \\ -114.414447331822828 \\ 55.7632172774061418 \\ 142.374104840575768 \\ 169.367713992202766 \end{bmatrix}$$

```
> A:=RandomMatrix(5,5,outputoptions=[shape=symmetric]);
```

$$A := \begin{bmatrix} 25 & 23 & -69 & 59 & -82 \\ 23 & -42 & 1 & 65 & 24 \\ -69 & 1 & -41 & -98 & -2 \\ 59 & 65 & -98 & -99 & -81 \\ -82 & 24 & -2 & -81 & 22 \end{bmatrix}$$

```
> A[1,1]:=1.0;   w:=Eigenvalues(A); # Not positive definite
```

$$A_{1,1} := 1.0$$

$$w := \begin{bmatrix} -216.601003534360756 \\ -86.3021609490597541 \\ -43.6657683700710280 \\ 13.7807456689493720 \\ 173.788187184542153 \end{bmatrix}$$

None of the three randomly generated matrices had all positive eigenvalues and so none of them are positive definite.

For symmetric matrices the fact that the eigenvector matrix P can be chosen to satisfy $P^T P = I$ means that the columns of P are orthonormal, that is they are orthogonal and have a length of one. Orthogonal matrices have very nice properties. One of the questions that we now want to study is this:

Question

Given a set of n linearly independent vectors S find an orthonormal set of n vectors Q such that span(S) = span(Q).

We can then replace the basis S of span(S) with the basis Q which is orthonormal. As we will see later this will be of great advantage, especially when it comes to expanding a given element of span(Q) as a linear combination of the basis elements.

5. Orthonormal vectors and the Gram-Schmidt Process

The Gram-Schmidt process is an algorithm that will create the set Q from the set S. In fact once the algorithm is complete we will have A decomposed as A = QR where $Q^T Q = I$. To see how the process works set

$S = \{v_1, v_2, \ldots, v_n\}$ and $Q = \{u_1, u_2, \ldots, u_n\}$. **Then set**

$$u_1 = v_1 / \|v_1\|_2$$

$$w = v_k - \sum_{j=1}^{k-1} r_{jk} u_j \quad k = 2,3,\ldots,n$$

$$u_m^T w = 0 = u_m^T v_k - r_{mk} \Rightarrow r_{mk} = u_m^T v_k \quad m = 1,2,\ldots,k-1$$

$$u_k = w / \|w\|_2$$

$$r_{kk} = \|w\|_2$$

To implement this in maple w need to write it as an algorithm that can be implemented. We assume that the vectors v are available as well as an appropriately sized matrix R which has been filled with zeros.

Algorithm (Gram-Schmidt)

1). $u_1 = v_1 / \|v_1\|_2$

2). $R[1,1] = \|v_1\|_2$

3). For k from 2 to n

4). For j from 1 to k-1

5) $R[j,k] = u_j^T v_k$

6). End for j

7). $w = v_k - \sum_{j-1}^{k-1} R[j,k]u_j$

8). $u_k = w/\|w\|_2$

9). $R[k,k] = \|w\|_2$

10). End for k

We shall write up our algorithm as a maple procedure and apply it to an example.

```
>restart; with(LinearAlgebra):
```

Gram-Schmidt procedure.

```
>GS:=proc(S::Matrix,R::Matrix) local j,k,m,n,w,w1,Q,u,z;
>w:=Column(S,1);
>R[1,1]:=evalf(Norm(w,2));
>w1:=map(evalf,evalm(w/R[1,1]));
>Q:=<<w1>>;
>n:=ColumnDimension(S);
>for k from 2 to n do
> for j from 1 to k-1 do
>  R[j,k]:=Transpose(Column(Q,j).Column(S,k));
>  u:=Vector(k-1,i->R[i,k]);
> od:
> w:=Column(S,k)-Q.u;
> z:=w/Norm(w,2);
> R[k,k]:=Norm(w,2);
> Q:=<Q|z>;
>od:
>Q;
>end:
```

The above procedure would be used on many occasions and so we really should write it up as a text file which we can read into any program on an as needed basis. Also the text file form is much easier to modify when adding or deleting features. Now lets apply the procedure to a matrix.

```
>m:=5: S:=RandomMatrix(m,m);  Rank(S);
```

$$S := \begin{bmatrix} -66 & -65 & 20 & -90 & 30 \\ 55 & 5 & -7 & -21 & 62 \\ 68 & 66 & 16 & -56 & -79 \\ 26 & -36 & -34 & -8 & -71 \\ 13 & -41 & -62 & -50 & 28 \end{bmatrix}$$

5

```
> R:=Matrix(m,m,0):
> Q:=GS(S,R); print(R);
```

$Q :=$

[-0.5822270406 , -0.308971411423570708 , 0.582092890062633428 ,
-0.475760633971869574 , -0.0191840228074152929]
[0.4851892005 , -0.325738833155357676 , 0.485257491361633820 ,
0.186364075674841644 , 0.623120574502632030]
[0.5998702842 , 0.306831635871208930 , 0.237166814655619286 ,
-0.630996924983675100 , -0.302662803076335862]
[0.2293621675 , -0.609055092046227208 , 0.127933552342101962 ,
0.299036989548182084 , -0.686042839013270900]
[0.1146810838 , -0.577353550432793906 , -0.594203620348921824 ,
-0.501338219864465695 , 0.221569793486195349]

[113.3578405 , 66.9031799329000024 , -20.3514815588999980 ,
1.04977299829999992 , -47.8484767849000008]
[0 , 84.3028143897998348 , 57.5138430339603844 , 51.2054891736056845 ,
-26.6276371090009008]
[0 , 0 , 44.5306081190283224 , -47.1733964462367226 , 3.09158922244732892]
[0 , 0 , 0 , 96.9152543599104490 , 11.8614143322684598]
[0 , 0 , 0 , 0 , 116.881312188268040]

[113.3578405 , 66.9031799329000024 , -20.3514815588999980 ,
1.04977299829999992 , -47.8484767849000008]
[0 , 84.3028143897998348 , 57.5138430339603844 , 51.2054891736056845 ,
-26.6276371090009008]
[0 , 0 , 44.5306081190283224 , -47.1733964462367226 , 3.09158922244732892]
[0 , 0 , 0 , 96.9152543599104490 , 11.8614143322684598]
[0 , 0 , 0 , 0 , 116.881312188268040]

We should check two things to make sure that our procedure is operating correctly. First we check that $Q^T Q = I$ and then we check that A – QR is the zero matrix.

```
> print(Transpose(Q).Q);
```

[0.99999999981569266, 0.146267303824387796 10^{-9},
-0.273145228657512007 10^{-9}, -0.208237642407294034 10^{-9},
-0.137714596670779344 10^{-10}]
[0.146267303824387796 10^{-9}, 1.00000000061868044,
-0.732214178178480779 10^{-9}, -0.684870715694785304 10^{-9},
0.289694795929662518 10^{-9}]
[-0.273145228657512007 10^{-9}, -0.732214178178480779 10^{-9},
0.99999999980858300, 0.296654312226252160 10^{-9},
-0.303672864632176243 10^{-9}]
[-0.208237642407294034 10^{-9}, -0.684870715694785304 10^{-9},
0.296654312226252160 10^{-9}, 1.00000000069321792,
-0.319469561915752818 10^{-9}]
[-0.137714596670779344 10^{-10}, 0.289694795929662518 10^{-9},
-0.303672864632176243 10^{-9}, -0.319469561915752818 10^{-9},
0.99999999981249454]

```
> print(S-Q.R);
```

[0.312181214212614577 10^{-8}, 0.805744093668181450 10^{-8},
0.248085996190638980 10^{-8}, 0.159816124778444646 10^{-7},
-0.210217621088304441 10^{-9}]
[-0.260151722386581241 10^{-8}, 0.849470183084122254 10^{-8},
0.206814476655381442 10^{-8}, -0.626028295869218710 10^{-8},
0.682813805497062276 10^{-8}]
[0.296672908461914630 10^{-8}, -0.800163491021521622 10^{-8},
0.101079500325340632 10^{-8}, 0.211962642993057671 10^{-7},
-0.331655769514327403 10^{-8}]
[-0.199282368384956498 10^{-9}, 0.158830957275313268 10^{-7},
0.545249179140228080 10^{-9}, -0.100451593709749432 10^{-7},
-0.751762740947015118 10^{-8}]
[-0.576753400594043342 10^{-8}, 0.150563792544744501 10^{-7},
-0.253247378623200348 10^{-8}, 0.168407865430708626 10^{-7},
0.242795294980169274 10^{-8}]

Again since we are using floating point numbers, which may be approximations only and not exact, we do not get the identity matrix as we should. Since the off diagonal elements are very small (in ten digit arithmetic) we may treat them as zeroes. Also A – QR is not the zero matrix but the entries in the difference are either zero or small and hence we may interpret them as zero.

6. QR factorization

Maple also has a function QRDecomposition **which decomposes A into A = QR. You should read the help page for it. The resulting Q will supply an orthonormal basis for the column space of A and if A is singular an orthonormal basis for the null space of A if we find .** $A^T = QR$ **.**

Example 6.1

Given the matrix

$$A = \begin{bmatrix} 1 & 0 & 0 & -1 & 0 & -1 \\ 2 & 1 & -1 & -2 & 1 & -1 \\ 5 & 4 & -3 & -6 & 4 & -1 \\ -5 & 0 & 4 & 1 & 0 & 5 \\ 9 & -8 & 15 & -16 & -8 & -17 \\ -9 & -5 & -4 & 18 & -5 & 4 \end{bmatrix}$$

1). Perform the QR decomposition and print out R

2). Show that the first three columns of Q are a basis for the column space of A.

3). Show that if we find the QR decomposition of the transpose of A then the last three columns of Q will be a basis for the null space (kernel) of A.

•

Solution

```
> restart; with(LinearAlgebra):
> A:=Matrix([[1,0,0,-1,0,-1],[2,1,-1,-2,1,-1],[5,4,-3,-6,4,-
1],[-5,0,4,1,0,5],[9,-8,15,-16,-8,-17],[-9,-5,-4,18,-
5,4]]);
```

$$A := \begin{bmatrix} 1 & 0 & 0 & -1 & 0 & -1 \\ 2 & 1 & -1 & -2 & 1 & -1 \\ 5 & 4 & -3 & -6 & 4 & -1 \\ -5 & 0 & 4 & 1 & 0 & 5 \\ 9 & -8 & 15 & -16 & -8 & -17 \\ -9 & -5 & -4 & 18 & -5 & 4 \end{bmatrix}$$

```
> Rank(A);
```

```
> w:=ColumnSpace(A);
```

$$w := \left[\begin{bmatrix} 1 \\ 0 \\ 0 \\ 7 \\ 46 \\ -26 \end{bmatrix}, \begin{bmatrix} 0 \\ 1 \\ 0 \\ -16 \\ -36 \\ 31 \end{bmatrix}, \begin{bmatrix} 0 \\ 0 \\ 1 \\ 4 \\ 7 \\ -9 \end{bmatrix}\right]$$

```
> C:=<w[1]|w[2]|w[3]>;
```

$$C := \begin{bmatrix} 1 & 0 & 0 \\ 0 & 1 & 0 \\ 0 & 0 & 1 \\ 7 & -16 & 4 \\ 46 & -36 & 7 \\ -26 & 31 & -9 \end{bmatrix}$$

Question #1

```
> (Q,R):=QRDecomposition(A,fullspan);
```

$Q, R :=$

$$\left[\frac{\sqrt{217}}{217}, \frac{5\sqrt{554001}}{1662003}, -\frac{593\sqrt{7779990}}{23339970}, \frac{3\sqrt{285969933510}}{1612070}, 0, 0\right]$$

$$\left[\frac{2\sqrt{217}}{217}, \frac{227\sqrt{554001}}{1662003}, -\frac{574\sqrt{7779990}}{11669985}, -\frac{15449\sqrt{285969933510}}{428954900265}, \frac{\sqrt{269631328638}}{532179}, 0\right]$$

$$\left[\frac{5\sqrt{217}}{217}, \frac{893\sqrt{554001}}{1662003}, -\frac{907\sqrt{7779990}}{11669985}, -\frac{10559\sqrt{285969933510}}{142984966755}, -\frac{62393\sqrt{269631328638}}{269631328638}, \frac{1003\sqrt{1519966}}{1519966}\right]$$

$$\left[-\frac{5\sqrt{217}}{217}, -\frac{25\sqrt{554001}}{1662003}, \frac{6961\sqrt{7779990}}{23339970}, \frac{132221\sqrt{285969933510}}{857909800530}, \frac{18581\sqrt{269631328638}}{57778141851}, \frac{35\sqrt{1519966}}{108569}\right]$$

$$\left[\frac{9\sqrt{217}}{217}, -\frac{1691\sqrt{554001}}{1662003}, \frac{676\sqrt{7779990}}{11669985}, -\frac{22834\sqrt{285969933510}}{428954900265}, \frac{24995\sqrt{269631328638}}{808893985914}, \frac{199\sqrt{1519966}}{1519966}\right]$$

$$\left[-\frac{9\sqrt{217}}{217}, -\frac{1130\sqrt{554001}}{1662003}, -\frac{1922\sqrt{7779990}}{11669985}, \frac{2701\sqrt{285969933510}}{142984966755},\right.$$

$$\frac{19043\sqrt{269631328638}}{134815664319}, \frac{242\sqrt{1519966}}{759983}\Bigg], \begin{bmatrix} \sqrt{217} & -\frac{5\sqrt{217}}{217} & \frac{134\sqrt{217}}{217} & -\cdot \\ 0 & \frac{3\sqrt{554001}}{217} & -\frac{1037\sqrt{554001}}{72261} & \frac{38}{} \\ 0 & 0 & \frac{\sqrt{7779990}}{333} & - \\ 0 & 0 & 0 & \\ 0 & 0 & 0 & \\ 0 & 0 & 0 & \end{bmatrix}$$

Question #2

```
> Q1:=<Column(Q,1)|Column(Q,2)|Column(Q,3)>;
```

$$Q1 := \begin{bmatrix} \frac{\sqrt{217}}{217} & \frac{5\sqrt{554001}}{1662003} & -\frac{593\sqrt{7779990}}{23339970} \\ \frac{2\sqrt{217}}{217} & \frac{227\sqrt{554001}}{1662003} & -\frac{574\sqrt{7779990}}{11669985} \\ \frac{5\sqrt{217}}{217} & \frac{893\sqrt{554001}}{1662003} & -\frac{907\sqrt{7779990}}{11669985} \\ -\frac{5\sqrt{217}}{217} & -\frac{25\sqrt{554001}}{1662003} & \frac{6961\sqrt{7779990}}{23339970} \\ \frac{9\sqrt{217}}{217} & -\frac{1691\sqrt{554001}}{1662003} & \frac{676\sqrt{7779990}}{11669985} \\ -\frac{9\sqrt{217}}{217} & -\frac{1130\sqrt{554001}}{1662003} & -\frac{1922\sqrt{7779990}}{11669985} \end{bmatrix}$$

```
> x:=Vector([x1,x2,x3]); w:=C.x;
```

$$x := \begin{bmatrix} x1 \\ x2 \\ x3 \end{bmatrix}$$

$$w := \begin{bmatrix} x1 \\ x2 \\ x3 \\ 7\,x1 - 16\,x2 + 4\,x3 \\ 46\,x1 - 36\,x2 + 7\,x3 \\ -26\,x1 + 31\,x2 - 9\,x3 \end{bmatrix}$$

```
> y:=LinearSolve(Q1,w);# showing that w is in the span of Q1
```

$$y := \begin{bmatrix} -\frac{\sqrt{217}\,(521\,x2 - 614\,x1 - 129\,x3)}{217} \\ \frac{\sqrt{2553}\,\sqrt{217}\,(1151\,x2 - 2112\,x1 - 38\,x3)}{72261} \\ -\frac{\sqrt{1612070}\,\sqrt{2553}\,(4\,x2 - 3\,x1 - x3)}{7659} \end{bmatrix}$$

```
> w:=Q1.x;
```

$$w := \begin{bmatrix} \frac{\sqrt{217}\,x1}{217} + \frac{5\sqrt{554001}\,x2}{1662003} - \frac{593\sqrt{7779990}\,x3}{23339970} \\ \frac{2\sqrt{217}\,x1}{217} + \frac{227\sqrt{554001}\,x2}{1662003} - \frac{574\sqrt{7779990}\,x3}{11669985} \\ \frac{5\sqrt{217}\,x1}{217} + \frac{893\sqrt{554001}\,x2}{1662003} - \frac{907\sqrt{7779990}\,x3}{11669985} \\ -\frac{5\sqrt{217}\,x1}{217} - \frac{25\sqrt{554001}\,x2}{1662003} + \frac{6961\sqrt{7779990}\,x3}{23339970} \\ \frac{9\sqrt{217}\,x1}{217} - \frac{1691\sqrt{554001}\,x2}{1662003} + \frac{676\sqrt{7779990}\,x3}{11669985} \\ -\frac{9\sqrt{217}\,x1}{217} - \frac{1130\sqrt{554001}\,x2}{1662003} - \frac{1922\sqrt{7779990}\,x3}{11669985} \end{bmatrix}$$

```
> b:=LinearSolve(C,w); # showing that w is in the span of C
```

$$b := \begin{bmatrix} \frac{\sqrt{2553}\,\sqrt{217}\,(210270\sqrt{2553}\,x1 + 350450\,x2 - 593\sqrt{217}\,\sqrt{1612070}\,x3)}{116489790270} \\ \frac{\sqrt{2553}\,\sqrt{217}\,(210270\sqrt{2553}\,x1 + 7955215\,x2 - 574\sqrt{217}\,\sqrt{1612070}\,x3)}{58244895135} \\ \frac{\sqrt{2553}\,\sqrt{217}\,(525675\sqrt{2553}\,x1 + 31295185\,x2 - 907\sqrt{217}\,\sqrt{1612070}\,x3)}{58244895135} \end{bmatrix}$$

Question #3

```
> w:=NullSpace(A);
```

$$w := \left\{ \begin{bmatrix} 1 \\ 1 \\ 1 \\ 1 \\ 0 \\ 0 \end{bmatrix}, \begin{bmatrix} 0 \\ -1 \\ 0 \\ 0 \\ 1 \\ 0 \end{bmatrix}, \begin{bmatrix} 1 \\ -1 \\ 0 \\ 0 \\ 0 \\ 1 \end{bmatrix} \right\}$$

```
> C:=<w[1]|w[2]|w[3]>;
```

$$C := \begin{bmatrix} 1 & 0 & 1 \\ -1 & -1 & 1 \\ 0 & 0 & 1 \\ 0 & 0 & 1 \\ 0 & 1 & 0 \\ 1 & 0 & 0 \end{bmatrix}$$

```
> A.C;
```

$$\begin{bmatrix} 0 & 0 & 0 \\ 0 & 0 & 0 \\ 0 & 0 & 0 \\ 0 & 0 & 0 \\ 0 & 0 & 0 \\ 0 & 0 & 0 \end{bmatrix}$$

The kernel **command returns a basis for the null space of A. Now we shall take the last three columns of Q and put them into a matrix Q1. There are three vectors and they are linearly independent because they are orthogonal. To show that it is a basis we need only show that A.Q1 is the zero matrix.**

```
> (Q,R):=QRDecomposition(Transpose(A),fullspan):
> Q1:=<Column(Q,4)|Column(Q,5)|Column(Q,6)>:
> evalm(A.Q1);
```

$$\begin{bmatrix} 0 & 0 & 0 \\ 0 & 0 & 0 \\ 0 & 0 & 0 \\ 0 & 0 & 0 \\ 0 & 0 & 0 \\ 0 & 0 & 0 \end{bmatrix}$$

7. Singular Value Decomposition.

The singular value decomposition provides a great deal of useful information about the matrix A. The maple command to perform the decomposition is **SingularValues.** Let A be an m by n matrix. Let U be an m by m orthonormal matrix and V an n by n orthonormal matrix. Let S be an m by n diagonal matrix with diagonal elements $s_1 \geq s_2 \geq \ldots \geq s_n \geq 0$. Then the singular value decomposition of A is $A = USV^T$.

Furthermore

1). If $p \leq n$ of the singular values (the s's) are nonzero then the first p columns of U provide an orthonormal basis for the column space of A.

2). The first p columns of V form an orthonormal basis for the row space of A.

3). The last n-p columns of V form an orthonormal basis for the null space of A.

4). The last m-p columns of U form an orthonormal basis for the null space of A^T.

5). The Euclidean (2) norm of A is given by the largest value among the s's.

We can solve the matrix equation Ax = B using

$$USV^T x = b$$
$$Sy = U^T b.$$
$$x = Vy$$

Example 7.1

Generate a 5 by 5 random matrix A and vector b of length 5 and the use the singular value decomposition to solve the matrix equation Ax = b.

```
> A:=RandomMatrix(5,5); U:=Copy(A):   V:=Copy(A):
b:=RandomVector(5):
```

$$A := \begin{bmatrix} -67 & -84 & -82 & 85 & -26 \\ -96 & -31 & -18 & -95 & 46 \\ -58 & -2 & 31 & 3 & 6 \\ -35 & 13 & 69 & 1 & -53 \\ 25 & -57 & -38 & 6 & -69 \end{bmatrix}$$

```
> (U,S,V):=SingularValues(A,output=['U','S','Vt']);
```

$U, S, V :=$

[-0.906882706260265904 , 0.136199287706755240 , 0.139845315903518952 , 0.317277891146194868 , -0.196955676546518688]
[0.0352528188728404857 , 0.925201890121145842 , -0.161257435469740029 , -0.311921124727643884 , -0.139498934519739908]
[0.0153322279271204398 , 0.279461294294940376 , 0.445707412803094394 , 0.234733373339768986 , 0.817258498512369314]
[0.137328059051388008 , 0.00245030016548159618 , 0.869212992173006848 , -0.244448074645167990 , -0.407245519639193198]
[-0.396518502508450110 , -0.217592733253038518 , -0.0159060848604695542 , -0.828966425123037420 , 0.328615492767872552],
$\begin{bmatrix} 173.985430638841308 \\ 155.628910405553028 \\ 103.038033631749741 \\ 78.1917177732711508 \\ 6.62805162659606495 \end{bmatrix}$,

[0.240066979775247158 , 0.571550528860202300 , 0.567567462670022226 ,
-0.474923870314079344 , 0.260791738568196263]
[-0.769002798531041476 , -0.181497650909021091 , -0.0688887488065017818 ,
-0.493365674300595192 , 0.357124558488669640]
[-0.490693168032350635 , 0.0443231951579946479 , 0.638913351657873996 ,
0.284528012710543622 , -0.517773317460114124]
[-0.218644696510133658 , 0.340470925608996056 , 0.0192902111854723196 ,
0.666145757818278228 , 0.626220218929472882]
[0.249834450401429514 , -0.722844104119325315 , 0.514333678103922387 ,
0.0795692663753723994 , 0.379748269185155472]

```
> SS:=DiagonalMatrix(S);
```

$$SS := \begin{bmatrix} 173.985430638841308\ ,0\ ,0\ ,0\ ,0 \\ 0\ ,155.628910405553028\ ,0\ ,0\ ,0 \\ 0\ ,0\ ,103.038033631749741\ ,0\ ,0 \\ 0\ ,0\ ,0\ ,78.1917177732711508\ ,0 \\ 0\ ,0\ ,0\ ,0\ ,6.62805162659606495 \end{bmatrix}$$

```
> y:=BackwardSubstitute(<SS|Transpose(U).b>);
```

$$y := \begin{bmatrix} 0.527484399149827364 \\ 0.472771795879168733 \\ -0.413055942938394316 \\ 0.838897787695258800 \\ 5.09581918305920832 \end{bmatrix}$$

```
> x:=Transpose(V).y;
```

$$x := \begin{bmatrix} 1.05544311445087002 \\ -3.20049348810310796 \\ 2.64004130647243152 \\ 0.363008501672602957 \\ 2.98073460938017832 \end{bmatrix}$$

```
> check:=A.x-b;
```

$$check := \begin{bmatrix} 0.568434188608080148\ 10^{-13} \\ -0.284217094304040074\ 10^{-13} \\ 0.213162820728030056\ 10^{-13} \\ 0.710542735760100186\ 10^{-13} \\ -0.198951966012828052\ 10^{-12} \end{bmatrix}$$

Example 7.2
Same as in problem 7.1 but use a size of 10 by 10 and suppress most of the printout.

Solution:

```
A:=RandomMatrix(10,10); U:=Copy(A):  V:=Copy(A):
b:=RandomVector(10):
```

$$A := \begin{bmatrix} 49 & -7 & -10 & 89 & -99 & 67 & 30 & 32 & -14 & -60 \\ -31 & -5 & 18 & 32 & 32 & -8 & 10 & -58 & -74 & 56 \\ -35 & 42 & 92 & -68 & 8 & -22 & -96 & 44 & -68 & -28 \\ -62 & 63 & -33 & -35 & 93 & 86 & -6 & -85 & 92 & -71 \\ 46 & 64 & 47 & 27 & 52 & -43 & 1 & 61 & -92 & 21 \\ 79 & -98 & -85 & 96 & -99 & 88 & 19 & 62 & 97 & 56 \\ -69 & 18 & 88 & -29 & 45 & 92 & -20 & 50 & -19 & -14 \\ -24 & -37 & -70 & 36 & -56 & -2 & 76 & -61 & -78 & -67 \\ 46 & -73 & -88 & -44 & -37 & -87 & 87 & -22 & 61 & -53 \\ -83 & -42 & -64 & 17 & 21 & -49 & 83 & -26 & 23 & -20 \end{bmatrix}$$

```
> (U,S,V):=SingularValues(A,output=['U','S','Vt']):
> SS:=DiagonalMatrix(S):
> y:=BackwardSubstitute(<SS|Transpose(U).b>):
> x:=Transpose(V).y:   print(Transpose(x));
```

[-0.518849626221197014, 2.41849737634667550, -1.94756753134402416,
-1.36513640415429282, -0.927237338248452004, 1.34716154821349888,
1.27759474719010080, 0.504922018497003688, -0.866208221411588486,
1.42156673914521758]

```
> check:=Transpose(A.x-b);
```

check := [$0.284217094304040075\ 10^{-12}$, $-0.284217094304040074\ 10^{-13}$,
$-0.255795384873636068\ 10^{-12}$, $0.255795384873636068\ 10^{-12}$,
$0.206057393370429054\ 10^{-12}$, $0.106581410364015028\ 10^{-12}$,
$-0.710542735760100186\ 10^{-14}$, $-0.284217094304040074\ 10^{-13}$,
$0.170530256582424045\ 10^{-12}$, $-0.106581410364015028\ 10^{-13}$]

Exercises

1). Generate random symmetric matrices of size 3 by 3, 5 by 5, 7 by 7 and 9 by 9. For each matrix answer the following questions.

a). find all of the eigenvalues and eigenvectors in floating point form.

b). Form the matrix P of eigenvectors and the diagonal matrix D containing the eigenvalues.

c). determine how close to the zero matrix AP – PD is.

d). determine how close to the zero matrix $A - PDP^T$ is.

2). For each of the matrices listed below answer the following questions.

$$\begin{bmatrix} 19 & 17 & 12 \\ -36 & -34 & -26 \\ 21 & 21 & 18 \end{bmatrix}$$

$$\begin{bmatrix} -48 & 25 & 8 & 9 & 12 \\ -73 & 38 & 14 & 15 & 18 \\ 50 & -26 & -12 & -12 & -12 \\ 1 & -1 & 4 & 4 & 0 \\ -62 & 32 & 4 & 6 & 16 \end{bmatrix}$$

$$\begin{bmatrix} -24 & -17 & 3 & 4 & 5 & 6 & 7 \\ 44 & 31 & -4 & -6 & -8 & -10 & -12 \\ 10 & 6 & 1 & -2 & -2 & -2 & -2 \\ 32 & 20 & -2 & 0 & -6 & -8 & -10 \\ -43 & -25 & 5 & 6 & 12 & 8 & 9 \\ 16 & 9 & -2 & -2 & -2 & 4 & -2 \\ -5 & -3 & 1 & 0 & -1 & -2 & 4 \end{bmatrix}$$

$$\begin{bmatrix} 21 & 16 & 3 & 4 & 5 & 7 & 8 & 9 & 10 \\ -44 & -31 & -4 & -6 & -8 & -12 & -14 & -16 & -18 \\ 4 & 4 & -1 & 2 & 2 & 2 & 2 & 2 & 2 \\ 42 & 26 & 2 & 2 & 6 & 10 & 12 & 14 & 16 \\ -66 & -44 & -5 & -8 & -12 & -17 & -20 & -23 & -26 \\ 68 & 48 & 5 & 8 & 11 & 18 & 20 & 23 & 26 \\ -13 & -15 & -5 & -6 & -7 & -9 & -8 & -11 & -12 \\ 34 & 20 & -1 & 0 & 1 & 3 & 4 & 8 & 6 \\ -37 & -22 & 1 & 0 & -1 & -3 & -4 & -5 & -2 \end{bmatrix}$$

2). For each matrix answer the following questions.

a). find all of the eigenvalues and eigenvectors in rational form.

b). Form the matrix P of eigenvectors and the diagonal matrix D containing the eigenvalues.

c). determine how close to the zero matrix AP – PD is.

d). determine how close to the zero matrix $A - PDP^T$ is.

Chapter 5
Linear Transformations using the linalg Package

Let $A = \left[a_{ij}\right]_{m \times n}$, $x^T = (x_1, x_2, \ldots, x_n)$ and $y^T = (y_1, y_2, \ldots, y_m)$. Then y = Ax is a function from R^n to R^m. This can also be expressed in the more usual notation y = f(x). In this case the function or transformation is generated by the matrix A. As usual R^n is the domain of the function and its range is $R(A) = \{b \in R^m : \text{for some } x \in R^n,\ b = Ax\}$. Some questions that come to mind immediately are:

1). Given a particular matrix A how can we more explicitly characterize its range?

2). Given that y = Ax is there anything that we can say about x?

3). Finally is A the only matrix that will generate the function?

1. Linear Independence and Span.

Let y = Ax and note that if $A = [c_1, c_2, \ldots, c_n]$, where $c_k, k = 1, 2, \ldots, n,$ are the columns of the matrix A then $y = \sum_{k=1}^{n} x_k c_k$. The sum is called a linear combination of the c_k's. This shows that y is a vector in the span of the columns of the matrix A. Recall the definition for the span of a set of vectors.

Definition 5.1

Let $S = \{v_1, v_2, \ldots, v_n\}$ be a set of vectors then the span of S is

$$\text{span(S)} = \left\{ \sum_{k=1}^{n} c_k v_k : c_k, k = 1, 2, \ldots, n \text{ are constants} \right\}.$$

Returning to the function y = Ax the next question to be answered is what is the minimal set, $S = \{c_1, c_2, \ldots, c_n\}$, of columns of the matrix A that will span the range? By this we mean the following. Suppose that $\sum_{k=1}^{n} x_k c_k = 0$ and not all of the x_k's are zero. To be more specific suppose that $x_r \neq 0$. Then $c_r = \sum_{k=1}^{r-1} \frac{x_k}{x_r} c_k + \sum_{k=r+1}^{n} \frac{x_k}{x_r} c_k$ and hence we can throw out c_r from S because it is a linear combination of the remaining columns. Continuing in this fashion we will soon reach a point where no more columns can be thrown out of the set S. This will be our minimal set. We need to recall our definition of linear independence.

Definition 5.2

The set of vectors $S = \{v_1, v_2, \ldots, v_n\}$ is linearly independent if $\sum_{k=1}^{n} c_k v_k = 0$, where $c_k, k = 1, 2, \ldots, n$ are constants, implies that all of the c_k's are zero. Otherwise they are linearly dependent.

Returning to y = Ax we see that what we want to do is to continue throwing out columns from the set S until it becomes a linearly independent set. At this point we will be unable to throw out any more column vectors and still have a set that spans R(A).

How many vectors will there be in the resulting linearly independent set? We can determine this by computing the rank of the matrix A. The rank of A is the number of linearly independent columns in A. This is always the same as the number of linearly independent rows in A. We can actually find a spanning set for the range of A by using the maple command **colspace(A).** To determine a spanning set for the null space of A, that is the set of vectors such that Ax = 0, we use the maple command **kernel(A).**

Example 5.1

Given the matrix

```
> A:=randmatrix(4,6); rank(A);
```

$$A := \begin{bmatrix} -5 & -28 & 4 & -11 & 10 & 57 \\ -82 & -48 & -11 & 38 & -7 & 58 \\ -94 & -68 & 14 & -35 & -14 & -9 \\ -51 & -73 & -73 & -91 & 1 & 5 \end{bmatrix}$$

$$4$$

1). Find a linearly independent set of vectors that span the column space of A.

2). Find a linearly independent set of vectors that span the null space of A.

3). Find a set of columns of A that span the range of A.

Solution:

1). The fact that the rank is 4 indicates that the spanning set should contain 4 vectors. The vectors that we need are

```
> colspace(A);
```

$$\{[1, 0, 0, 0], [0, 1, 0, 0], [0, 0, 1, 0], [0, 0, 0, 1]\}$$

2). The spanning set of vectors for the null space should contain 6-4 = 2 vectors. One such set is

```
> kernel(A);
```

$$\left\{\left[\frac{-3496547}{3875756}, 1, \frac{289349}{7751512}, \frac{-295110}{968939}, \frac{15494979}{7751512}, 0\right],\right.$$
$$\left.\left[\frac{930247}{968939}, 0, \frac{-44867}{1937878}, \frac{-511781}{968939}, \frac{-11223629}{1937878}, 1\right]\right\}$$

3). Now we wish to find not just a spanning set for the column space of A but one that consists only of columns of the original matrix A. The easiest way to find them is to generate the reduced row echelon form using the **gaussjord** command. Then select those columns from A that correspond to columns of the reduced row echelon form which contain leading ones. The reduced row echelon form is

```
> B:=gaussjord(A);
```

$$B := \begin{bmatrix} 1 & 0 & 0 & 0 & \frac{6993094}{15494979} & \frac{25625750}{15494979} \\ 0 & 1 & 0 & 0 & \frac{-7751512}{15494979} & \frac{-44894516}{15494979} \\ 0 & 0 & 1 & 0 & \frac{-289349}{15494979} & \frac{-1317076}{15494979} \\ 0 & 0 & 0 & 1 & \frac{786960}{5164993} & \frac{7285927}{5164993} \end{bmatrix}$$

Notice that the first four columns of B contain leading ones and so we select the first 4 columns from A.

```
> C:=augment(col(A,1),col(A,2),col(A,3),col(A,4));
```

$$C := \begin{bmatrix} -5 & -28 & 4 & -11 \\ -82 & -48 & -11 & 38 \\ -94 & -68 & 14 & -35 \\ -51 & -73 & -73 & -91 \end{bmatrix}$$

To check that these are linearly independent we use.

```
> rank(C);
```

$$4$$

Since the rank of the original matrix is also 4 this tells us that these columns are linearly independent and are thus a spanning set for the range of A.

2. Subspaces and Basis

In section 2 we are going to clarify some of the results in section 1. We will do this by refreshing our memory on the ideas of subspace and basis.

Definition 5.3

A set of vectors $S \subseteq R^n$ is a subspace of R^n if it contains all vectors of the form $\left\{ \sum_{k=1}^{s} c_k v_k : v_k \in S, c_k \text{ are constants } k = 1, 2, \ldots s \right\}$.

The definition is not easy to apply and so we have the following theorem.

Theorem 5.1

$S \subseteq R^n$ is a subspace if

1). $v \in S$ and c is a constant $\Rightarrow$ $cv \in S$.

2). $v_1, v_2 \in S \Rightarrow v_1 + v_2 \in S.$

Number 1 says that S is closed under scalar multiplication and 2 says that it is closed under addition.

Exercise 5.2

Show that $S = \{(a,0,b,0,c) : a, b,$ and c are constants$\}$ is a subspace of R^5.

Solution:

Problems of this kind are done pretty much the same way. We will show that 1 and 2 of theorem 5.1 are satisfied. First we need to identify the main characteristics of the set S. Examining the set S we see that the identifying characteristics of S are that elements 2 and 4 of a vector in S are zero while the other three positions are filled with any constant.

1). Let v = (a,0,b,0,c) and k be a constant then kv = (ka,0,kb,0,kc) belongs to S since positions 2 and 4 are zero.

2). Let v = (a,0,b,0,c) and w = (d,0,e,0,f) then v + w = (a+d,0,b+e,0,c+f) belongs to S since positions 2 and 4 are zero..

Since 1 and 2 hold by the theorem S is a subspace.

Exercise 5.3

Show that the set $S = \{(x, y) : 3x - 2y = 0\}$ is a subspace of R^2.

Solution:

We proceed in the same way as we did in Example 5.2.

1). Let v = (a,b) belong to S and let k be a constant then kv = (ka,kb) and so 3ka – 2kb = k(3a -2 b) = 0. Thus kv is in S.

2). Let v = (a,b) and w = (c,d) so that v + w = (a+c,b+d) and so 3(a+c) - 2(b+d) = (3a-2b) + (3c-2d) =0. Thus v + w is in S.

Thus 1 and 2 are satisfied and so S is a subspace.

Next we turn our attention to basis.

Definition 5.4

Let S be a subspace of R^n and $T = \{v_1, v_2, \ldots, v_r\} \subseteq S$ then T is a basis for S if

1). T is a linearly independent set of vectors.

2). Span(T) = S.

It can be shown that given a subspace S, every basis for S contains exactly the same number of vectors. If we have a basis for a subspace S then we have all the information that we need to know about S because if $y \in S$ then $y = \sum_{k=1}^{r} c_k v_k$ for some choice of the constants. It can be shown that the choice for the c_k's is unique.

Given y = Ax we would like to find a basis for the column space of A and the kernel of A. As we saw in section 1 these are given by the maple commands **colspace(A)** and **kernel(A)**.

Exercise 5.4

Generate a random matrix of size 5 by 5 and then

1). Find any basis for the range of A.

2). Find any basis for the kernel of A.

3). Find a basis for the range of A which consists of columns of the original matrix A.

Solution:

```
> A:=matrix(5,5,(i,j)->i+j);    rank(A);
```

$$A := \begin{bmatrix} 2 & 3 & 4 & 5 & 6 \\ 3 & 4 & 5 & 6 & 7 \\ 4 & 5 & 6 & 7 & 8 \\ 5 & 6 & 7 & 8 & 9 \\ 6 & 7 & 8 & 9 & 10 \end{bmatrix}$$

$$2$$

The fact that the rank of A is two tells us that the a basis for the column space will contain 2 vectors while the basis for the null space will contain $5 - 2 = 3$ vectors.

Question #1

```
> w:=colspace(A);
```

$$w := \{[1, 0, -1, -2, -3], [0, 1, 2, 3, 4]\}$$

The two vectors in w form a basis for the column space of A.

Question #2

```
> q:=kernel(A);
```

$$q := \{[1, -2, 1, 0, 0], [2, -3, 0, 1, 0], [3, -4, 0, 0, 1]\}$$

The kernel command returns a basis for the null space of A.

Question #3

First we find the reduced row echelon form of A to determine where the leading ones are.

```
> B:=gaussjord(A);
```

$$B := \begin{bmatrix} 1 & 0 & -1 & -2 & -3 \\ 0 & 1 & 2 & 3 & 4 \\ 0 & 0 & 0 & 0 & 0 \\ 0 & 0 & 0 & 0 & 0 \\ 0 & 0 & 0 & 0 & 0 \end{bmatrix}$$

Since the leading ones are in column one and two we select columns 1 and 2 of A as the basis.

```
> C:=augment(col(A,1),col(A,2));   rank(C);
```

$$C := \begin{bmatrix} 2 & 3 \\ 3 & 4 \\ 4 & 5 \\ 5 & 6 \\ 6 & 7 \end{bmatrix}$$

$$2$$

The fact that the rank is two shows that the columns of the matrix C are linearly independent and since there are two of them they form a basis for the column space of A.

Exercise 5.5

Find a basis for the subspace of Exercise 5.3 $S = \{(x, y): 3x - 2y = 0\}$.

Solution:

Solving 3x – 2y = 0 for y gives y = 3x/2. From this we see that the coordinates of any point on this line will have the form (2a,3a) = a(2,3). Hence a basis for this subspace contains the one single vector (2,3).

3. Linear Transformations

In the preceding sections we have considered the function y = Ax. We have concentrated on finding a basis for the column space of A, the kernel of A and finding a basis for the column space of A that consists of columns of A itself. Now we want to concentrate on the vector x. In particular given y we want to find the vector x that maps into it. Unless the range is all of R^m not every y is going to be in the range of A.

Given y then we wish to answer two questions in this section.

1). Is the given vector y in the range of A? That is, is there a vector x such that y = Ax?

2). If the answer to question 1 is yes then what is x?

This is exactly the same as the problem of solving Ax = b. If b is not in the range of A then the system of equations is inconsistent. We can recognize this situation by finding the row echelon form of the augmented matrix [A|b}. If any row consists of all zeros except for the rightmost element and this is nonzero then the system is inconsistent. Another way of saying this is that b is not in the range of A.

If the system is consistent the solution for x can be represented as x = u + v where Au = b and Aw = 0. If the solution is unique then w = 0. Otherwise the solution is not unique and this will always be the case when A is m by n where m<n.

For some examples we have

Exercise 5.6

Given the matrix $A = \begin{bmatrix} 1 & 1 & 0 & -1 \\ 0 & 2 & 2 & 2 \\ -1 & 0 & 1 & 2 \\ 2 & 5 & 3 & 1 \end{bmatrix}$ and $b_1^T = [2,2,-1,7]$ and $b_2^T = [2,0,-1,3]$.

Determine if b_1 and b_2 are in the range of A. For each one that is in the range of A find all of the x's that map into it.

Solution:

```
> A:=matrix([[1,1,0,-1],[0,2,2,2],[-1,0,1,2],[2,5,3,1]]);
```

$$A := \begin{bmatrix} 1 & 1 & 0 & -1 \\ 0 & 2 & 2 & 2 \\ -1 & 0 & 1 & 2 \\ 2 & 5 & 3 & 1 \end{bmatrix}$$

```
> b1:=vector([2,2,-1,7]);    b2:=vector([2,0,-1,3]);
```

$$b1 := [2, 2, -1, 7]$$

$$b2 := [2, 0, -1, 3]$$

Is b1 in the range of A?

```
> B1:=augment(A,b1): B11:=gausselim(B1); #b1 is in the range
of A
```

$$B11 := \begin{bmatrix} 1 & 1 & 0 & -1 & 2 \\ 0 & 2 & 2 & 2 & 2 \\ 0 & 0 & 0 & 0 & 0 \\ 0 & 0 & 0 & 0 & 0 \end{bmatrix}$$

```
> x:=backsub(B11);
```

$$x := [1 + \_t_2 + 2\,\_t_1,\ 1 - \_t_2 - \_t_1,\ \_t_2,\ \_t_1]$$

The matrix B11 shows that b1 is in the range of A because the system of equations is consistent. Now x can be written as x = [1,1,0,0] + _t[1][2,-1,0,1] + _t[2][1,-1,1,0]. Thus The vector [1,1,0,0] maps into b1 while the other two vectors constitute a basis for the null space of A.

Is b2 in the range of A?

```
> B2:=augment(A,b2): B21:=gausselim(B2); # b2 is not in the
range of A
```

$$B21 := \begin{bmatrix} 1 & 1 & 0 & -1 & 2 \\ 0 & 2 & 2 & 2 & 0 \\ 0 & 0 & 0 & 0 & 1 \\ 0 & 0 & 0 & 0 & 0 \end{bmatrix}$$

Row 3 of B21 shows us that b2 is not in the range of A because the system of equations is inconsistent.

How did we find b2 in the above example so that it would not be in the range of A? Well one way of doing this is to start with a set S that consists of the columns of A that form a basis for the column space of A and then fill it out to a basis for R^n. In the above example n = 4. How can we fill out S to a basis for the whole space? One way to do this is to form the augmented matrix [S|I], where I is the n by n identity matrix. Then find the reduced row echelon form. Finally take the columns of [S|I] that correspond to leading ones in the reduced row echelon form.

Exercise 5.7

Given the matrix $A = \begin{bmatrix} 1 & 1 & 0 & -1 \\ 0 & 2 & 2 & 2 \\ -1 & 0 & 1 & 2 \\ 2 & 5 & 3 & 1 \end{bmatrix}$ find:

1). A basis S for the column space of A consisting of columns of A.

2). Fill this basis out to a basis for the whole space.

3). Use the results of 2 to find all b's that are not in the range of A.

Solution:

```
> A:=matrix([[1,1,0,-1],[0,2,2,2],[-1,0,1,2],[2,5,3,1]]);
```

$$A := \begin{bmatrix} 1 & 1 & 0 & -1 \\ 0 & 2 & 2 & 2 \\ -1 & 0 & 1 & 2 \\ 2 & 5 & 3 & 1 \end{bmatrix}$$

Question #1

```
> A1:=gaussjord(A);
```

$$A1 := \begin{bmatrix} 1 & 0 & -1 & -2 \\ 0 & 1 & 1 & 1 \\ 0 & 0 & 0 & 0 \\ 0 & 0 & 0 & 0 \end{bmatrix}$$

A1 shows that we should take the first two columns of A as a basis for the column space of A.

```
> S:=augment(col(A,1),col(A,2));
```

$$S := \begin{bmatrix} 1 & 1 \\ 0 & 2 \\ -1 & 0 \\ 2 & 5 \end{bmatrix}$$

Question #2

```
> I1:=diag(1,1,1,1):
> Q:=augment(S,I1): Q11:=gaussjord(Q);
```

$$Q11 := \begin{bmatrix} 1 & 0 & 0 & 0 & -1 & 0 \\ 0 & 1 & 0 & 0 & \frac{2}{5} & \frac{1}{5} \\ 0 & 0 & 1 & 0 & \frac{3}{5} & \frac{-1}{5} \\ 0 & 0 & 0 & 1 & \frac{-4}{5} & \frac{-2}{5} \end{bmatrix}$$

Q11 shows that we should take columns 3 and 4 from Q and add them to S to form a basis for the whole space.

```
> T:=augment(S,col(Q,3),col(Q,4));
```

$$T := \begin{bmatrix} 1 & 1 & 1 & 0 \\ 0 & 2 & 0 & 1 \\ -1 & 0 & 0 & 0 \\ 2 & 5 & 0 & 0 \end{bmatrix}$$

Next we show that any linear combination of the last two columns of T will fail to be in the range of A.

Question #3

```
> v:=vector([a,b,0,0]);
```

$$v := [a, b, 0, 0]$$

```
> P:=augment(A,v):   P1:=gaussjord(P);
```

$$P1 := \begin{bmatrix} 1 & 0 & -1 & -2 & 0 \\ 0 & 1 & 1 & 1 & 0 \\ 0 & 0 & 0 & 0 & 1 \\ 0 & 0 & 0 & 0 & 0 \end{bmatrix}$$

This shows that for any a and b the vector is not in the range of A.

4. Change of Basis

Let e_j, $j = 1,2,\ldots,n$ be a vector of length n such that every component of e_j is zero except for the j th component which is a one. More formally

Definition 5.5

Let $\delta_{ij} = \begin{cases} 0 & \text{if } i \neq j \\ 1 & \text{if } i = j \end{cases}$ then $e_j = (\delta_{1j}, \delta_{2j,}, \ldots, \delta_{nj})$, for all $j = 1,2,\ldots,n$.

If $S = \{e_1, e_2, \ldots, e_n\}$ then S is called the standard basis for R^n.

As an example the standard basis for R^3 is $e_1 = (1,0,0)$, $e_2 = (0,1,0$, $e_3 = (0,0,1)$. Every vector in R^3 can be represented as a liner combination of these basis elements.

Example 5.8

Given the vector v = (x,y,z) represent v as a linear combination of the standard basis.

Solution:

$$\begin{aligned} v &= (x, y, z) \\ &= x(1,0,0) + y(0,1,0) + z(0,0,1) \\ &= xe_1 + ye_2 + ze_3 \end{aligned}$$

Given the function y = Ax, where A is m by n, what is the underlying basis that is being used for the vectors x and y? The basis for both R^n and R^m is the standard basis. Now suppose that we change the basis for R^m. Let us change the basis from the standard basis to $S = \{v_1, v_2, \ldots, v_m\}$. How will this change the matrix A? Note that the underlying function y = f(x) does not change only its representation by A. If we use the standard basis for both R^n and R^m then we can compute y = f(x) as y = Ax.. Changing the basis for R^m does not effect the underlying function y = f(x) however it may very well change the matrix A in $y = Ax$. When referring to the standard basis we shall use x and y, when referring to a different basis we shall use y_M and x_N. Taking S as the new basis for R^m gives $y_M = (y_1, y_2, \ldots, y_m)^T = \sum_{k=1}^{m} y_k v_k$. The next question is how does the expression for A change? Since S is a basis each e_j can be written as a linear combination of the new basis elements. Thus

$$e_k = \sum_{j=1}^{m} r_{jk} v_j \text{ and so}$$

$$y = \sum_{k=1}^{m} y_k e_k$$

$$\begin{aligned} y_M &= \sum_{k=1}^{m} y_k \sum_{j=1}^{m} r_{jk} v_j \\ &= \sum_{j=1}^{m} \left(\sum_{k=1}^{m} r_{jk} y_k \right) v_j \\ &= \left(\sum_{k=1}^{m} r_{1k} y_k, \sum_{k=1}^{m} r_{2k} y_k, \ldots, \sum_{k=1}^{m} r_{mk} y_k \right) \end{aligned}$$

Where the last result is interpreted as the coefficient multipliers of the vectors v_k. If we set $R = \left[r_{ij} \right]_{m \times m}$ then the new representation for our function y = f(x) is
$y_M = (RA)x$ or $y_M = A_M x$.

The above results can be expressed in matrix form as follows. Let S be the matrix with columns $v_1, v_2, \ldots, v_m$ then the matrix R satisfies SR = I. From this we obtain
$y_M = (RA)x = (S^{-1}A)x$.

Example 5.9

Let $A = \begin{bmatrix} 1 & 2 & -1 & 0 \\ -1 & 0 & 2 & 1 \\ 0 & 1 & 3 & -1 \end{bmatrix}$ and $S = \left\{ v_1^T = [1,1,1], v_2^T = [1,1,0], v_3^T = [1,0,0] \right\}$. Using the standard basis a function has the representation y = Ax find the new representation if the basis for the range space is changed to S and apply this in particular to $x^T = [1,2,3,4]$..

Solution:

```
> A:=matrix([[1,2,-1,0],[-1,0,2,1],[0,1,3,-1]]);
S:=matrix([[1,1,1],[1,1,0],[1,0,0]]);
```

$$A := \begin{bmatrix} 1 & 2 & -1 & 0 \\ -1 & 0 & 2 & 1 \\ 0 & 1 & 3 & -1 \end{bmatrix}$$

$$S := \begin{bmatrix} 1 & 1 & 1 \\ 1 & 1 & 0 \\ 1 & 0 & 0 \end{bmatrix}$$

Note that the new basis vectors are now the columns of S. Next we expand each member of the standard basis in terms of the new basis vectors contained in S.

```
> e1:=vector([1,0,0]);   e2:=vector([0,1,0]);
e3:=vector([0,0,1]);
```

$$e1 := [1, 0, 0]$$

$$e2 := [0, 1, 0]$$

$$e3 := [0, 0, 1]$$

```
> r1:=linsolve(S,e1);
```

$$r1 := [0, 0, 1]$$

```
> r2:=linsolve(S,e2);
```

$$r2 := [0, 1, -1]$$

```
> r3:=linsolve(S,e3);
```

$$r3 := [1, -1, 0]$$

```
> R:=augment(r1,r2,r3);
```

$$R := \begin{bmatrix} 0 & 0 & 1 \\ 0 & 1 & -1 \\ 1 & -1 & 0 \end{bmatrix}$$

Next we check our work to by testing to see that SR = I where I is the 3 by 3 identity matrix.

```
> B:=evalm(S&*R);
```

$$B := \begin{bmatrix} 1 & 0 & 0 \\ 0 & 1 & 0 \\ 0 & 0 & 1 \end{bmatrix}$$

```
> x:=vector([1,2,3,4]); # referred to the standard basis.
```

$$x := [1, 2, 3, 4]$$

```
> y:=evalm(A&*x); # y is referred to the standard basis.
```

$$y := [2, 9, 7]$$

```
> z:=evalm(R&*y);# z is y referred to the basis vectors in S
```

$$z := [7, 2, -7]$$

Finally we check our work by showing that Sz is y referred to the standard basis.

```
> q:=evalm(S&*z);
```

$$q := [2, 9, 7]$$

Now it is time to consider the representation of y = f(x) when we change the basis for R^n. Let $T = \{u_1, u_2, \ldots, u_n\}$ be a new basis for R^n. We shall also denote by T the matrix which has for its columns $u_1, u_2, \ldots, u_n$. We have $e_k = \sum_{j=1}^{n} p_{jk} u_j$ and so

$$x = \sum_{k=1}^{n} x_k e_k$$

$$x_N = \sum_{k=1}^{n} x_k \sum_{j=1}^{n} p_{jk} u_j$$

$$= \sum_{j=1}^{n}\left(\sum_{k=1}^{n} p_{jk} x_k\right) u_j$$

$$= \left(\sum_{k=1}^{n} p_{1k} x_k, \sum_{k=1}^{n} p_{2k} x_k, \ldots, \sum_{k=1}^{n} p_{mk} x_k\right)$$

In terms of matrix operations this can be expressed as TP = I and from there $x_N = Px$ $y = AP^{-1}x_N = ATx_N$. Finally putting the two results together gives $y_M = \left(S^{-1}AT\right)x_N$.

Exercise 5.10

Let $A = \begin{bmatrix} 1 & 2 & -1 & 0 \\ -1 & 0 & 2 & 1 \\ 0 & 1 & 3 & -1 \end{bmatrix}$ and $S = \{v_1^T = [1,1,1], v_2^T = [1,1,0], v_3^T = [1,0,0]\}$ and let

$T = \{v_1 = [1,-1,1,-1], v_2 = [1,0,1,1], v_3 = [0,2,3,-1], v_4 = [1,0,0,2]\}$. Using the standard basis a function has the representation y = Ax find the new representation if the basis for the range space is changed to S, the basis for the domain is changed to T and apply this in particular to $x^T = [1,2,3,4]$.

Solution:

```
> restart; with(linalg):
Warning, the protected names norm and trace have been redefined and
unprotected
```

```
> A:=matrix([[1,2,-1,0],[-1,0,2,1],[0,1,3,-1]]);
S:=matrix([[1,1,1],[1,1,0],[1,0,0]]);
```

$$A := \begin{bmatrix} 1 & 2 & -1 & 0 \\ -1 & 0 & 2 & 1 \\ 0 & 1 & 3 & -1 \end{bmatrix}$$

$$S := \begin{bmatrix} 1 & 1 & 1 \\ 1 & 1 & 0 \\ 1 & 0 & 0 \end{bmatrix}$$

```
> T:=matrix([[1,1,0,1],[-1,0,2,0],[1,1,3,0],[-1,1,-1,2]]);
x:=vector([1,2,3,4]);
```

$$T := \begin{bmatrix} 1 & 1 & 0 & 1 \\ -1 & 0 & 2 & 0 \\ 1 & 1 & 3 & 0 \\ -1 & 1 & -1 & 2 \end{bmatrix}$$

$$x := [1, 2, 3, 4]$$

First we compute y where y and x are referred to the standard basis.

```
>y:=evalm(A&*x); # referred to the standard basis
```

$$y := [2, 9, 7]$$

Compute the matrix B so that y = f(x) is ym = Bxn in the basis S and T.

```
>B:=evalm(inverse(S)&*A&*T);
```

$$B := \begin{bmatrix} 3 & 2 & 12 & -2 \\ -3 & 0 & -7 & 3 \\ -2 & -2 & -4 & 0 \end{bmatrix}$$

```
>xn:=evalm(inverse(T)&*x);
```

$$xn := \left[-3, \frac{15}{2}, \frac{-1}{2}, \frac{-7}{2} \right]$$

recompute y where y and x are referred to the new basis S and T.

```
>ym:=evalm(B&*xn); # referred to the basis S and T.
```

$$ym := [7, 2, -7]$$

We now check to see if our results are correct by expanding ym in the standard basis.

```
>y:=evalm(S&*ym);
```

$$y := [2, 9, 7]$$

As a conclusion we see that the change of basis results are best expressed as matrix equations involving the new basis vectors, that is in terms of the matrixes S and T.

Exercises for chapter 5

1). Determine if the columns of the following matrices are linearly independent or not. If they are not linearly independent then determine a set of columns from the original matrices that are a basis for the column space.

$$A = \begin{bmatrix} 1 & 1 & 1 & 0 & 1 \\ 2 & 3 & 2 & -1 & 2 \\ -1 & 1 & -1 & -2 & -1 \end{bmatrix}, \quad B = \begin{bmatrix} 1 & -1 & -1 & 0 \\ -1 & 2 & 0 & -1 \\ 2 & -4 & 1 & 3 \\ 2 & -6 & 3 & 5 \end{bmatrix}, \quad C = \begin{bmatrix} 1 & -1 & -1 & 1 \\ -2 & 3 & 2 & -3 \\ 5 & -4 & -5 & 4 \\ 2 & 0 & -2 & 0 \end{bmatrix}.$$

2). Find an orthonormal basis for the span of the column space of

$$S = \begin{bmatrix} 1 & -1 & -1 & -1 \\ 5 & -4 & -5 & -5 \\ 4 & 1 & -3 & -4 \\ 1 & 3 & 2 & -1 \end{bmatrix}.$$

3). Determine if the following subsets of R^3 are subspaces or not. Justify your answer.

a). $S = \{(x, y, z) : x + 3y = 5z,\ x, y, z \in R\}$.

b). $S = \{(x, y, z) : x + 3y = 5,\ x, y, z \in R\}$.

c). $S = \{(x, 0, z) : x = 3z,\ x, y, z \in R\}$.

d). $S = \{(x, y, z) : xyz = 0,\ x, y, z \in R\}$.

4). For those S's, in (3), that are subspaces find a basis for the subspace.

5). For the matrices in problem 1 find:

a). A basis for the range space.

b). A basis for the null space.

6). Let y = f(x) be a function represented by y = Ax where

$$A = \begin{bmatrix} 1 & 1 & 1 & 0 & 1 \\ 2 & 3 & 2 & -1 & 2 \\ -1 & 1 & -1 & -2 & -1 \end{bmatrix}$$

and x and y are referred to the standard basis. Let the columns of

$$S = \begin{bmatrix} 1 & 0 & 1 \\ -4 & 1 & -3 \\ -4 & -2 & -5 \end{bmatrix}$$

be a new basis for the range of f and let

$$T = \begin{bmatrix} 1 & 1 & -1 & 0 & 1 \\ -4 & -3 & 4 & 1 & -5 \\ 3 & -1 & -2 & -3 & 6 \\ -4 & -6 & 2 & -4 & 0 \\ -2 & -7 & 7 & 0 & -2 \end{bmatrix}$$

be a new basis for the domain of f. Find the new representation of y = f(x) as y = Bx

where now x and y are referred to the new basis.

Project 1.

Given
$$S1 = \{v_1 = [1,2,3,4], v_2 = [1,0,1,-1], v_3 = [2,0,-3,0], v_4 = [1,-1,1,-1]\},$$
$$S2 = \{v_1 = [1,0,0,0], v_2 = [1,1,0,0], v_3 = [1,1,1,0], v_4 = [1,1,1,1]\},$$
$$S3 = \{v_1 = [0,1,-1,0], v_2 = [1,0,0,2], v_3 = [1,2,-2,1], v_4 = [0,1,1,2]\},$$
$$S4 = \{v_1 = [4,-3,0,1], v_2 = [2,0,0,4], v_3 = [1,-2,-1,3], v_4 = [0,5,6,0]\}$$

1). Show that each of the sets S1, S2, S3, and S4 are linearly independent and that they span R^4.

2). Given v = [1,-5,2,-7] express v as a linear combination v_k, k = 1, 2, 3, 4, for each of the sets S1, S2, S3, and S4.

3). Drop vector v_4 from the set S1 and replace it with a vector that will make the set S1 linearly dependent. Note that the solution to this problem is not unique.

4). Let A1 be the matrix with columns v_k, k = 1, 2, 3, 4, from S1 in (1). Show that $A1^{-1}$ has rank 4 and hence its columns are linearly independent. Show that any vector $u \in R^4$ can be represented as a linear combination of the columns of $A1^{-1}$.

5). Is it possible to create a new linearly independent set S5 by selecting exactly one vector from each of the sets S1, S2, S3, and S4? If so find such a set.

Project 2.

Given

$$S1=\{v_1=[1,2,0,-1]\},$$
$$S2=\{v_1=[1,0,1,2],v_2=[1,1,1,0]\},$$
$$S3=\{v_1=[1,1,-1,0],v_2=[1,0,1,2],v_3=[1,2,-2,1]\},$$
$$S4=\{v_1=[0,2,0,1],v_2=[2,1,-1,4],v_3=[1,-2,-1,3],v_4=[-1,-2,3,2]\}$$

along with

$$T=\{u_1=[3,6,9,10],u_2=[5,9,1,2],u_3=[1,-1,0,1]\}$$

1). Show that the sets S1, S2, S3, and S4 are linearly independent.

2). Test each vector u_1, u_2, and u_3 to see if it is in the span of Sk, k = 1, 2, 3, 4. If it is in the span then represent it as a linear combination of the vectors in Sk.

3). Which of the sets S1, S2, S3, and S4, if any, are a basis for R^4? Justify your answer by expanding an arbitrary element of R^4 in terms of the basis elements.

4). For each of the sets S1, S2, S3, and S4 that fails to be a basis fill it out to a basis. That is find a basis for R^4 that includes the vectors in Sk. Then represent each vector in the set T as a linear combination of the basis elements.

5). Is it possible to create a new basis by selecting exactly one vector from each of the sets Sk, k = 1, 2, 3, 4? If so find such a basis.

Project 3.

Given $A = \begin{bmatrix} 1 & -1 & -1 & -1 & -1 \\ 3 & -3 & -4 & -4 & -2 \\ -5 & 5 & 3 & 3 & 7 \\ 5 & -5 & -1 & -1 & -9 \end{bmatrix}$

1). Find a basis for the column space of A that consists of columns of the original matrix A.

2). Find a basis for R^4 that contains the basis found in (1).

3). Using the results from (2) characterize all those vectors in R^4 that do not belong to the span of the column space of A.

4). Generate five random vectors of length 4 and determine which, if any, are in the column space of A.

5). Characterize the span of A and use this characterization to construct a vector that is not in the range of A.

6). Pick a column of the matrix A that is not in the basis found in (1) and modify it so that the resulting matrix has rank 3.

Project 4.

Given
$$S1=\{v_1=[1,2,3,4],v_2=[1,0,1,-1],v_3=[2,0,-3,0],v_4=[1,-1,1,-1]\},$$
$$S2=\{v_1=[1,0,0,0],v_2=[1,1,0,0],v_3=[1,1,1,0],v_4=[1,1,1,1]\},$$
$$S3=\{v_1=[0,1,-1,0],v_2=[1,0,0,2],v_3=[1,2,-2,1],v_4=[0,1,1,2]\},$$
$$S4=\{v_1=[4,-3,0,1],v_2=[2,0,0,4],v_3=[1,-2,-1,3],v_4=[0,5,6,0]\}$$

and let
$$A=\begin{bmatrix} 1 & -1 & 1 & -1 \\ 4 & -3 & 3 & -5 \\ -1 & 6 & -6 & -4 \\ 0 & 1 & -1 & -1 \end{bmatrix}$$

$T=\{u_1=[3,6,9,10],u_2=[5,9,1,2],u_3=[1,-1,0,1]\}$. Using the natural basis for R^4 a function y = f(x) can be represented by the matrix equation y = Ax. Under these conditions:

1). Determine the dimension of the range of A.

2). Take the basis for the range of f to be S1 and then represent each of the vectors $y_k=f(u_k),\ k=1,2,3$ in terms of the basis S1. Do the same with S1 replaced by S2, S3 or S4.

3). Take the basis for the range of f to be S1 and find the matrix representation for the function y = f(x). Repeat this procedure with the set S1 replaced by S2, S3, or S4.

4). Find the matrix representation for the function y = f(x) if the basis for the (domain(f), range(f)) is (S1,S2) or (S3,S4).

5). Take the basis for the domain(f) to be S2 and for the range(f) take the basis S4. First find the image $y_k=f(u_k),\ k=1,2,3$ when referred to the natural basis. Then find the image when the basis S2 and S4 are used. Check your results by expanding the y_k's in the natural basis to see if the result is the same.

Chapter 5.1

Linear Transformations using the LinearAlgebra Package

Let $A=\left[a_{ij}\right]_{m\times n}$, $x^T=(x_1,x_2,\ldots,x_n)$ and $y^T=(y_1,y_2,\ldots,y_m)$. Then y = Ax is a function from R^n to R^m. This can also be expressed in the more usual notation y = f(x). In this case the function or transformation is generated by the matrix A. As usual R^n is the domain of the function and its range is $R(A)=\{b\in R^m : \text{for some } x\in R^n,\ b=Ax\}$. Some questions that come to mind immediately are:

1). Given a particular matrix A how can we more explicitly characterize its range?

2). Given that y = Ax is there anything that we can say about x?

3). Finally is A the only matrix that will generate the function?

1. Linear Independence and Span.

Let y = Ax and note that if $A=[c_1,c_2,\ldots,c_n]$, where c_k, $k=1,2,\ldots,n$, are the columns of the matrix A then $y=\sum_{k=1}^{n}x_k c_k$. The sum is called a linear combination of the c_k's. This shows that y is a vector in the span of the columns of the matrix A. Recall the definition for the span of a set of vectors.

Definition 5.1

Let $S=\{v_1,v_2,\ldots,v_n\}$ be a set of vectors then the span of S is

$$\text{span}(S)=\left\{\sum_{k=1}^{n}c_k v_k : c_k,\ k=1,2,\ldots,n \text{ are constants}\right\}.$$

Returning to the function y = Ax the next question to be answered is what is the minimal set, $S=\{c_1,c_2,\ldots,c_n\}$, of columns of the matrix A that will span the range? By this we mean the following. Suppose that $\sum_{k=1}^{n}x_k c_k=0$ and not all of the x_k's are zero. To be more specific suppose that $x_r\neq 0$. Then $c_r=\sum_{k=1}^{r-1}\frac{x_k}{x_r}c_k+\sum_{k=r+1}^{n}\frac{x_k}{x_r}c_k$ and hence we can throw out c_r from S because it is a linear combination of the remaining columns. Continuing in this fashion we will soon reach a point where no more columns can be thrown out of the set S. This will be our minimal set. We need to recall our definition of linear independence.

Definition 5.2

The set of vectors $S = \{v_1, v_2, \ldots, v_n\}$ is linearly independent if $\sum_{k=1}^{n} c_k v_k = 0$, where $c_k, k = 1,2,\ldots,n$ are constants, implies that all of the c_k's are zero. Otherwise they are linearly dependent.

Returning to y = Ax we see that what we want to do is to continue throwing out columns from the set S until it becomes a linearly independent set. At this point we will be unable to throw out any more column vectors and still have a set that spans R(A).

How many vectors will there be in the resulting linearly independent set? We can determine this by computing the rank of the matrix A. The rank of A is the number of linearly independent columns in A. This is always the same as the number of linearly independent rows in A. We can actually find a spanning set for the range of A by using the maple command **ColumnSpace(A).** To determine a spanning set for the null space of A, that is the set of vectors such that Ax = 0, we use the maple command **NullSpace(A).**

Example 5.1

Given the matrix

```
> A:=RandomMatrix(4,6);   Rank(A);
```

$$A := \begin{bmatrix} 55 & -65 & -41 & -34 & -56 & 62 \\ 68 & 5 & 20 & -62 & -8 & -79 \\ 26 & 66 & -7 & -90 & -50 & -71 \\ 13 & -36 & 16 & -21 & 30 & 28 \end{bmatrix}$$

$$4$$

1). Find a linearly independent set of vectors that span the column space of A.

2). Find a linearly independent set of vectors that span the null space of A.

3). Find a set of columns of A that span the range of A.

Solution:

1). The fact that the rank is 4 indicates that the spanning set should contain 4 vectors. The vectors that we need are

```
> ColumnSpace(A);
```

$$\left[\begin{bmatrix} 1 \\ 0 \\ 0 \\ 0 \end{bmatrix}, \begin{bmatrix} 0 \\ 1 \\ 0 \\ 0 \end{bmatrix}, \begin{bmatrix} 0 \\ 0 \\ 1 \\ 0 \end{bmatrix}, \begin{bmatrix} 0 \\ 0 \\ 0 \\ 1 \end{bmatrix}\right]$$

2). The spanning set of vectors for the null space should contain 6-4 = 2 vectors. One such set is

```
> NullSpace(A);
```

$$\left\{\begin{bmatrix}\frac{6772546}{5320371}\\\frac{6922033}{5320371}\\\frac{4067579}{5320371}\\\frac{2519122}{5320371}\\0\\1\end{bmatrix}, \begin{bmatrix}\frac{2534486}{5320371}\\\frac{2168552}{5320371}\\\frac{-7247126}{5320371}\\\frac{-69640}{5320371}\\1\\0\end{bmatrix}\right\}$$

3). Now we wish to find not just a spanning set for the column space of A but one that consists only of columns of the original matrix A. The easiest way to find them is to generate the reduced row echelon form using the **ReducedRowEchelonForm** command. Then select those columns from A that correspond to columns of the reduced row echelon form which contain leading ones. The reduced row echelon form is

```
> B:=ReducedRowEchelonForm(A);
```

$$B := \begin{bmatrix}1 & 0 & 0 & 0 & \frac{-2534486}{5320371} & \frac{-6772546}{5320371}\\0 & 1 & 0 & 0 & \frac{-2168552}{5320371} & \frac{-6922033}{5320371}\\0 & 0 & 1 & 0 & \frac{7247126}{5320371} & \frac{-4067579}{5320371}\\0 & 0 & 0 & 1 & \frac{69640}{5320371} & \frac{-2519122}{5320371}\end{bmatrix}$$

Notice that the first four columns of B contain leading ones and so we select the first 4 columns from A.

```
> C:=<Column(A,1)|Column(A,2)|Column(A,3)|Column(A,4)>;
```

$$C := \begin{bmatrix}55 & -65 & -41 & -34\\68 & 5 & 20 & -62\\26 & 66 & -7 & -90\\13 & -36 & 16 & -21\end{bmatrix}$$

To check that these are linearly independent we use.

```
> Rank(C);
```

$$4$$

Since the rank of the original matrix is also 4 this tells us that these columns are linearly independent and are thus a spanning set for the range of A.

2. Subspaces and Basis

In section 2 we are going to clarify some of the results in section 1. We will do this by reviewing the ideas of subspace and basis.

Definition 5.3

A set of vectors $S \subseteq R^n$ is a subspace of R^n if it contains all vectors of the form

$$\left\{\sum_{k=1}^{s} c_k v_k : v_k \in S, c_k \text{ are constants } k = 1,2,\ldots s\right\}.$$

The definition is not easy to apply and so we have the following theorem.

Theorem 5.1

$S \subseteq R^n$ is a subspace if

1). $v \in S$ and c is a constant $\Rightarrow$ $cv \in S$.

2). $v_1, v_2 \in S \Rightarrow v_1 + v_2 \in S$.

Number 1 says that S is closed under scalar multiplication and 2 says that it is closed under addition.

Exercise 5.2

Show that $S = \{(a,0,b,0,c) : a,b,$ and c are constants$\}$ is a subspace of R^5.

Solution:

Problems of this kind are done pretty much the same way. We will show that 1 and 2 of theorem 5.1 are satisfied. First we need to identify the main characteristics of the set S. Examining the set S we see that the identifying characteristics of S are that elements 2 and 4 of a vector in S are zero while the other three positions are filled with any constant.

1). Let v = (a,0,b,0,c) and k be a constant then kv = (ka,0,kb,0,kc) belongs to S since positions 2 and 4 are zero.

2). Let v = (a,0,b,0,c) and w = (d,0,e,0,f) then v + w = (a+d,0,b+e,0,c+f) belongs to S since positions 2 and 4 are zero..

Since 1 and 2 hold by the theorem S is a subspace.

Exercise 5.3

Show that the set $S = \{(x,y) : 3x - 2y = 0\}$ is a subspace of R^2.

Solution:

We proceed in the same way as we did in Example 5.2.

1). Let v = (a,b) belong to S and let k be a constant then kv = (ka,kb) and so 3ka – 2kb = k(3a -2 b) = 0. Thus kv is in S.

2). Let v = (a,b) and w = (c,d) so that v + w = (a+c,b+d) and so 3(a+c) - 2(b+d) = (3a-2b) + (3c-2d) =0. Thus v + w is in S.

Thus 1 and 2 are satisfied and so S is a subspace.

Next we turn our attention to basis.

Definition 5.4

Let S be a subspace of R^n and $T = \{v_1, v_2, \ldots, v_r\} \subseteq S$ then T is a basis for S if

1). T is a linearly independent set of vectors.

2). Span(T) = S.

It can be shown that given a subspace S, every basis for S contains exactly the same number of vectors. If we have a basis for a subspace S then we have all the information that we need to know about S because if $y \in S$ then $y = \sum_{k=1}^{r} c_k v_k$ for some choice of the constants. It can be shown that the choice for the c_k's is unique.

Given y = Ax we would like to find a basis for the column space of A and the kernel of A. As we saw in section 1 these are given by the maple commands **ColumnSpace(A)** and **NullSpace(A)**.

Exercise 5.4

Generate a random matrix of size 5 by 5 and then

1). Find any basis for the range of A.

2). Find any basis for the kernel of A.

3). Find a basis for the range of A which consists of columns of the original matrix A.

Solution:

```
> A:=Matrix(5,5,(i,j)->i+j);    Rank(A);
```

$$A := \begin{bmatrix} 2 & 3 & 4 & 5 & 6 \\ 3 & 4 & 5 & 6 & 7 \\ 4 & 5 & 6 & 7 & 8 \\ 5 & 6 & 7 & 8 & 9 \\ 6 & 7 & 8 & 9 & 10 \end{bmatrix}$$

$$2$$

The fact that the rank of A is two tells us that the a basis for the column space will contain 2 vectors while the basis for the null space will contain $5 - 2 = 3$ vectors.

Question #1

```
> w:=ColumnSpace(A);
```

$$w := \left[\begin{bmatrix} 1 \\ 0 \\ -1 \\ -2 \\ -3 \end{bmatrix}, \begin{bmatrix} 0 \\ 1 \\ 2 \\ 3 \\ 4 \end{bmatrix}\right]$$

The two vectors in w form a basis for the column space of A.

Question #2

```
> q:=NullSpace(A);
```

$$q := \left\{\begin{bmatrix} 2 \\ -3 \\ 0 \\ 1 \\ 0 \end{bmatrix}, \begin{bmatrix} 1 \\ -2 \\ 1 \\ 0 \\ 0 \end{bmatrix}, \begin{bmatrix} 3 \\ -4 \\ 0 \\ 0 \\ 1 \end{bmatrix}\right\}$$

The **NullSpace** command returns a basis for the null space of A.

Question #3

First we find the reduced row echelon form of A to determine where the leading ones are.

```
> B:=ReducedRowEchelonForm(A);
```

$$B := \begin{bmatrix} 1 & 0 & -1 & -2 & -3 \\ 0 & 1 & 2 & 3 & 4 \\ 0 & 0 & 0 & 0 & 0 \\ 0 & 0 & 0 & 0 & 0 \\ 0 & 0 & 0 & 0 & 0 \end{bmatrix}$$

Since the leading ones are in column one and two we select columns 1 and 2 of A as the basis.

```
> C:=<Column(A,1)|Column(A,2)>;     Rank(C);
```

$$C := \begin{bmatrix} 2 & 3 \\ 3 & 4 \\ 4 & 5 \\ 5 & 6 \\ 6 & 7 \end{bmatrix}$$

$$2$$

The fact that the rank is two shows that the columns of the matrix C are linearly independent and since there are two of them they form a basis for the column space of A.

Exercise 5.5

Find a basis for the subspace of Exercise 5.3 $S = \{(x, y) : 3x - 2y = 0\}$.

Solution:

Solving 3x – 2y = 0 for y gives y = 3x/2. From this we see that the coordinates of any point on this line will have the form (2a,3a) = a(2,3). Hence a basis for this subspace contains the one single vector (2,3).

3. Linear Transformations

In the preceding sections we have considered the function y = Ax. We have concentrated on finding a basis for the column space of A, the kernel of A and finding a basis for the column space of A that consists of columns of A itself. Now we want to concentrate on the vector x. In particular given y we want to find the vector x that maps into it. Unless the range is all of R^m not every y is going to be in the range of A.

Given y then we wish to answer two questions in this section.

1). Is the given vector y in the range of A? That is, is there a vector x such that y = Ax?

2). If the answer to question 1 is yes then what is x?

This is exactly the same as the problem of solving Ax = b. If b is not in the range of A then the system of equations is inconsistent. We can recognize this situation by finding the row echelon form of the augmented matrix [A|b}. If any row consists of all zeros except for the rightmost element and this is nonzero then the system is inconsistent. Another way of saying this is that b is not in the range of A.

If the system is consistent the solution for x can be represented as x = u + v where Au = b and Aw = 0. If the solution is unique then w = 0. Otherwise the solution is not unique and this will always be the case when A is m by n where m<n.

For some examples we have

Exercise 5.6

Given the matrix $A = \begin{bmatrix} 1 & 1 & 0 & -1 \\ 0 & 2 & 2 & 2 \\ -1 & 0 & 1 & 2 \\ 2 & 5 & 3 & 1 \end{bmatrix}$ and $b_1^T = [2,2,-1,7]$ and $b_2^T = [2,0,-1,3]$.

Determine if b_1 and b_2 are in the range of A. For each one that is in the range of A find all of the x's that map into it.

Solution:

```
> A:=Matrix([[1,1,0,-1],[0,2,2,2],[-1,0,1,2],[2,5,3,1]]);
```

$$A := \begin{bmatrix} 1 & 1 & 0 & -1 \\ 0 & 2 & 2 & 2 \\ -1 & 0 & 1 & 2 \\ 2 & 5 & 3 & 1 \end{bmatrix}$$

```
> b1:=Vector([2,2,-1,7]);    b2:=Vector([2,0,-1,3]);
```

$$b1 := \begin{bmatrix} 2 \\ 2 \\ -1 \\ 7 \end{bmatrix}$$

$$b2 := \begin{bmatrix} 2 \\ 0 \\ -1 \\ 3 \end{bmatrix}$$

Is b1 in the range of A?

```
> B1:=<A|b1>: B11:=ReducedRowEchelonForm(B1); #b1 is in the
range of A
```

$$B11 := \begin{bmatrix} 1 & 0 & -1 & -2 & 1 \\ 0 & 1 & 1 & 1 & 1 \\ 0 & 0 & 0 & 0 & 0 \\ 0 & 0 & 0 & 0 & 0 \end{bmatrix}$$

```
> x:=BackwardSubstitute(B11);
```

$$x := \begin{bmatrix} 1 + \_t_2 + 2\,\_t_1 \\ 1 - \_t_2 - \_t_1 \\ \_t_2 \\ \_t_1 \end{bmatrix}$$

The matrix B11 shows that b1 is in the range of A because the system of equations is consistent. Now x can be written as x = [1,1,0,0] + _t[1][2,-1,0,1] + _t[2][1,-1,1,0]. Thus The vector [1,1,0,0] maps into b1 while the other two vectors constitute a basis for the null space of A.

Is b2 in the range of A?

```
> B2:=<A|b2>: B21:=ReducedRowEchelonForm(B2); # b2 is not in
the range of A
```

$$B21 := \begin{bmatrix} 1 & 0 & -1 & -2 & 0 \\ 0 & 1 & 1 & 1 & 0 \\ 0 & 0 & 0 & 0 & 1 \\ 0 & 0 & 0 & 0 & 0 \end{bmatrix}$$

Row 3 of B21 shows us that b2 is not in the range of A because the system of equations is inconsistent.

How did we find b2 in the above example so that it would not be in the range of A? Well one way of doing this is to start with a set S that consists of the columns of A that form a basis for the column space of A and then fill it out to a basis for R^n. In the above example n = 4. How can we fill out S to a basis for the whole space? One way to do this is to form the augmented matrix [S|I], where I is the n by n identity matrix. Then find the reduced row echelon form. Finally take the columns of [S|I] that correspond to leading ones in the reduced row echelon form.

Exercise 5.7

Given the matrix $A = \begin{bmatrix} 1 & 1 & 0 & -1 \\ 0 & 2 & 2 & 2 \\ -1 & 0 & 1 & 2 \\ 2 & 5 & 3 & 1 \end{bmatrix}$ find:

1). A basis S for the column space of A consisting of columns of A.

2). Fill this basis out to a basis for the whole space.

3). Use the results of 2 to find all b's that are not in the range of A.

Solution:

```
> A:=Matrix([[1,1,0,-1],[0,2,2,2],[-1,0,1,2],[2,5,3,1]]);
```

$$A := \begin{bmatrix} 1 & 1 & 0 & -1 \\ 0 & 2 & 2 & 2 \\ -1 & 0 & 1 & 2 \\ 2 & 5 & 3 & 1 \end{bmatrix}$$

Question #1

```
> A1:=ReducedRowEchelonForm(A);
```

$$A1 := \begin{bmatrix} 1 & 0 & -1 & -2 \\ 0 & 1 & 1 & 1 \\ 0 & 0 & 0 & 0 \\ 0 & 0 & 0 & 0 \end{bmatrix}$$

A1 shows that we should take the first two columns of A as a basis for the column space of A.

```
> S:=<Column(A,1)|Column(A,2)>;
```

$$S := \begin{bmatrix} 1 & 1 \\ 0 & 2 \\ -1 & 0 \\ 2 & 5 \end{bmatrix}$$

Question #2

```
> I1:=IdentityMatrix(4);
```

$$I1 := \begin{bmatrix} 1 & 0 & 0 & 0 \\ 0 & 1 & 0 & 0 \\ 0 & 0 & 1 & 0 \\ 0 & 0 & 0 & 1 \end{bmatrix}$$

```
> Q:=<S|I1>: Q11:=ReducedRowEchelonForm(Q);
```

$$Q11 := \begin{bmatrix} 1 & 0 & 0 & 0 & -1 & 0 \\ 0 & 1 & 0 & 0 & \frac{2}{5} & \frac{1}{5} \\ 0 & 0 & 1 & 0 & \frac{3}{5} & \frac{-1}{5} \\ 0 & 0 & 0 & 1 & \frac{-4}{5} & \frac{-2}{5} \end{bmatrix}$$

Q11 shows that we should take columns 3 and 4 from Q and add them to S to form a basis for the whole space.

```
> T:=<S|Column(Q,3)|Column(Q,4)>;
```

$$T := \begin{bmatrix} 1 & 1 & 1 & 0 \\ 0 & 2 & 0 & 1 \\ -1 & 0 & 0 & 0 \\ 2 & 5 & 0 & 0 \end{bmatrix}$$


```
> T:=<S|Column(Q,3)|Column(Q,4)>;
```

$$T := \begin{bmatrix} 1 & 1 & 1 & 0 \\ 0 & 2 & 0 & 1 \\ -1 & 0 & 0 & 0 \\ 2 & 5 & 0 & 0 \end{bmatrix}$$

Next we show that any linear combination of the last two columns of T will fail to be in the range of A.

Question #3

```
> v:=Vector([a,b,0,0]);
```

$$v := \begin{bmatrix} a \\ b \\ 0 \\ 0 \end{bmatrix}$$

```
> P:=<A|v>:  P1:=ReducedRowEchelonForm(P); # v is not in the range of A.
```

$$P1 := \begin{bmatrix} 1 & 0 & -1 & -2 & 0 \\ 0 & 1 & 1 & 1 & 0 \\ 0 & 0 & 0 & 0 & 1 \\ 0 & 0 & 0 & 0 & 0 \end{bmatrix}$$

This shows that for any a and b the vector is not in the range of A.

4. Change of Basis

Let e_j, $j = 1,2,\ldots,n$ be a vector of length n such that every component of e_j is zero except for the j th component which is a one. More formally

Definition 5.5

Let $\delta_{ij} = \begin{cases} 0 & \text{if } i \neq j \\ 1 & \text{if } i = j \end{cases}$ then $e_j = \left(\delta_{1j}, \delta_{2j,}, \ldots, \delta_{nj}\right)$, for all $j = 1,2,\ldots,n$.

If $S = \{e_1, e_2, \ldots, e_n\}$ then S is called the standard basis for R^n.

As an example the standard basis for R^3 is $e_1 = (1,0,0)$, $e_2 = (0,1,0$, $e_3 = (0,0,1)$. Every vector in R^3 can be represented as a liner combination of these basis elements.

Example 5.8

Given the vector v = (x,y,z) represent v as a linear combination of the standard basis.

Solution:

$$
\begin{aligned}
v &= (x, y, z) \\
&= x(1,0,0) + y(0,1,0) + z(0,0,1) \\
&= xe_1 + ye_2 + ze_3
\end{aligned}
$$

Given the function y = Ax, where A is m by n, what is the underlying basis that is being used for the vectors x and y? The basis for both R^n and R^m is the standard basis. Now suppose that we change the basis for R^m. Let us change the basis from the standard basis to $S = \{v_1, v_2, \ldots, v_m\}$. How will this change the matrix A? Note that the underlying function y = f(x) does not change only its representation by A. If we use the standard basis for both R^n and R^m then we can compute y = f(x) as y = Ax.. Changing the basis for R^m does not effect the underlying function y = f(x) however it may very well change the matrix A in $y = Ax$. When referring to the standard basis we shall use x and y, when referring to a different basis we shall use y_M and x_N. Taking S as the new basis for R^m gives $y_M = (y_1, y_2, \ldots, y_m)^T = \sum_{k=1}^{m} y_k v_k$. The next question is how does the expression for A change? Since S is a basis each e_j can be written as a linear combination of the new basis elements. Thus

$$e_k = \sum_{j=1}^{m} r_{jk} v_j \text{ and so}$$

$$\begin{aligned} y &= \sum_{k=1}^{m} y_k e_k \\ y_M &= \sum_{k=1}^{m} y_k \sum_{j=1}^{m} r_{jk} v_j \\ &= \sum_{j=1}^{m}\left(\sum_{k=1}^{m} r_{jk} y_k\right) v_j \\ &= \left(\sum_{k=1}^{m} r_{1k} y_k, \sum_{k=1}^{m} r_{2k} y_k, \ldots, \sum_{k=1}^{m} r_{mk} y_k\right) \end{aligned}$$

Where the last result is interpreted as the coefficient multipliers of the vectors v_k. If we set $R = \left[r_{ij}\right]_{m \times m}$ then the new representation for our function y = f(x) is $y_M = (RA)x$ or $y_M = A_M x$.

The above results can be expressed in matrix form as follows. Let S be the matrix with columns $v_1, v_2, \ldots, v_m$ then the matrix R satisfies SR = I. From this we obtain $y_M = (RA)x = (S^{-1}A)x$.

Example 5.9

Let $A = \begin{bmatrix} 1 & 2 & -1 & 0 \\ -1 & 0 & 2 & 1 \\ 0 & 1 & 3 & -1 \end{bmatrix}$ and $S = \left\{v_1^T = [1,1,1], v_2^T = [1,1,0], v_3^T = [1,0,0]\right\}$. Using the standard basis a function has the representation y = Ax find the new representation if the basis for the range space is changed to S and apply this in particular to $x^T = [1,2,3,4]$..

Solution:

```
> A:=Matrix([[1,2,-1,0],[-1,0,2,1],[0,1,3,-1]]);
S:=Matrix([[1,1,1],[1,1,0],[1,0,0]]);
```

$$A := \begin{bmatrix} 1 & 2 & -1 & 0 \\ -1 & 0 & 2 & 1 \\ 0 & 1 & 3 & -1 \end{bmatrix}$$

$$S := \begin{bmatrix} 1 & 1 & 1 \\ 1 & 1 & 0 \\ 1 & 0 & 0 \end{bmatrix}$$

Note that the new basis vectors are now the columns of S. Next we expand each member of the standard basis in terms of the new basis vectors contained in S.

```
> e1:=Vector([1,0,0]);
e2:=Vector([0,1,0]);e3:=Vector([0,0,1]);
```

$$e1 := \begin{bmatrix} 1 \\ 0 \\ 0 \end{bmatrix}$$

$$e2 := \begin{bmatrix} 0 \\ 1 \\ 0 \end{bmatrix}$$

$$e3 := \begin{bmatrix} 0 \\ 0 \\ 1 \end{bmatrix}$$

```
> r1:=LinearSolve(S,e1);
```

$$r1 := \begin{bmatrix} 0 \\ 0 \\ 1 \end{bmatrix}$$

```
> r2:=LinearSolve(S,e2);
```

$$r2 := \begin{bmatrix} 0 \\ 1 \\ -1 \end{bmatrix}$$

```
> r3:=LinearSolve(S,e3);
```

$$r3 := \begin{bmatrix} 1 \\ -1 \\ 0 \end{bmatrix}$$

```
> R:=<r1|r2|r3>;
```

$$R := \begin{bmatrix} 0 & 0 & 1 \\ 0 & 1 & -1 \\ 1 & -1 & 0 \end{bmatrix}$$

Next we check our work to by testing to see that SR = I where I is the 3 by 3 identity matrix.

```
> B:=S.R;
```

$$B := \begin{bmatrix} 1 & 0 & 0 \\ 0 & 1 & 0 \\ 0 & 0 & 1 \end{bmatrix}$$

```
> x:=Vector([1,2,3,4]); # referred to the standard basis.
```

$$x := \begin{bmatrix} 1 \\ 2 \\ 3 \\ 4 \end{bmatrix}$$

```
>y:=A.x; # y is referred to the standard basis.
```

$$y := \begin{bmatrix} 2 \\ 9 \\ 7 \end{bmatrix}$$

```
>z:=R.y; # z is y referred to the basis vectors in S
```

$$z := \begin{bmatrix} 7 \\ 2 \\ -7 \end{bmatrix}$$

Finally we check our work by showing that Sz is y referred to the standard basis.

```
>q:=S.z;
```

$$q := \begin{bmatrix} 2 \\ 9 \\ 7 \end{bmatrix}$$

Now it is time to consider the representation of y = f(x) when we change the basis for R^n. Let $T = \{u_1, u_2, \ldots, u_n\}$ be a new basis for R^n. We shall also denote by T the matrix which has for its columns $u_1, u_2, \ldots, u_n$. We have $e_k = \sum_{j=1}^{n} p_{jk} u_j$ and so

$$
\begin{aligned}
x &= \sum_{k=1}^{n} x_k e_k \\
x_N &= \sum_{k=1}^{n} x_k \sum_{j=1}^{n} p_{jk} u_j \\
&= \sum_{j=1}^{n} \left(\sum_{k=1}^{n} p_{jk} x_k \right) u_j \\
&= \left(\sum_{k=1}^{n} p_{1k} x_k, \sum_{k=1}^{n} p_{2k} x_k, \ldots, \sum_{k=1}^{n} p_{mk} x_k \right)
\end{aligned}
$$

In terms of matrix operations this can be expressed as TP = I and from there $x_N = Px$ $y = AP^{-1} x_N = ATx_N$. Finally putting the two results together gives $y_M = \left(S^{-1}AT\right) x_N$.

Exercise 5.10

Let $A = \begin{bmatrix} 1 & 2 & -1 & 0 \\ -1 & 0 & 2 & 1 \\ 0 & 1 & 3 & -1 \end{bmatrix}$ and $S = \{v_1^T = [1,1,1], v_2^T = [1,1,0], v_3^T = [1,0,0]\}$ and let $T = \{v_1 = [1,-1,1,-1], v_2 = [1,0,1,1], v_3 = [0,2,3,-1], v_4 = [1,0,0,2]\}$. Using the standard basis a function has the representation y = Ax find the new representation if the basis for the range space is changed to S, the basis for the domain is changed to T and apply this in particular to $x^T = [1,2,3,4]$.

Solution:

```
> restart; with(LinearAlgebra):
> A:=Matrix([[1,2,-1,0],[-1,0,2,1],[0,1,3,-1]]);
S:=Matrix([[1,1,1],[1,1,0],[1,0,0]]);
```

$$A := \begin{bmatrix} 1 & 2 & -1 & 0 \\ -1 & 0 & 2 & 1 \\ 0 & 1 & 3 & -1 \end{bmatrix}$$

$$S := \begin{bmatrix} 1 & 1 & 1 \\ 1 & 1 & 0 \\ 1 & 0 & 0 \end{bmatrix}$$

```
> T:=Matrix([[1,1,0,1],[-1,0,2,0],[1,1,3,0],[-1,1,-1,2]]);
x:=Vector([1,2,3,4]);
```

$$T := \begin{bmatrix} 1 & 1 & 0 & 1 \\ -1 & 0 & 2 & 0 \\ 1 & 1 & 3 & 0 \\ -1 & 1 & -1 & 2 \end{bmatrix}$$

$$x := \begin{bmatrix} 1 \\ 2 \\ 3 \\ 4 \end{bmatrix}$$

First we compute y where y and x are referred to the standard basis.

```
> y:=A.x;
```

$$y := \begin{bmatrix} 2 \\ 9 \\ 7 \end{bmatrix}$$

Compute the matrix B so that y = f(x) is ym = Bxn in the basis S and T.

```
> B:=MatrixInverse(S).A.T;
```

$$B := \begin{bmatrix} 3 & 2 & 12 & -2 \\ -3 & 0 & -7 & 3 \\ -2 & -2 & -4 & 0 \end{bmatrix}$$

```
> xn:=MatrixInverse(T).x;
```

$$xn := \begin{bmatrix} -3 \\ \frac{15}{2} \\ \frac{-1}{2} \\ \frac{-7}{2} \end{bmatrix}$$

recompute y where y and x are referred to the new basis S and T.

```
> ym:=B.xn;
```

$$ym := \begin{bmatrix} 7 \\ 2 \\ -7 \end{bmatrix}$$

We now check to see if our results are correct by expanding ym in the standard basis.

```
> y:=S.ym;
```

$$y := \begin{bmatrix} 2 \\ 9 \\ 7 \end{bmatrix}$$

As a conclusion we see that the change of basis results are best expressed as matrix equations involving the new basis vectors, that is in terms of the matrixes S and T.

Exercises for chapter 5

1). Determine if the columns of the following matrices are linearly independent or not. If they are not linearly independent then determine a set of columns from the original matrices that are a basis for the column space.

$$A = \begin{bmatrix} 1 & 1 & 1 & 0 & 1 \\ 2 & 3 & 2 & -1 & 2 \\ -1 & 1 & -1 & -2 & -1 \end{bmatrix}, \quad B = \begin{bmatrix} 1 & -1 & -1 & 0 \\ -1 & 2 & 0 & -1 \\ 2 & -4 & 1 & 3 \\ 2 & -6 & 3 & 5 \end{bmatrix}, \quad C = \begin{bmatrix} 1 & -1 & -1 & 1 \\ -2 & 3 & 2 & -3 \\ 5 & -4 & -5 & 4 \\ 2 & 0 & -2 & 0 \end{bmatrix}.$$

2). Find an orthonormal basis for the span of the column space of

$$S = \begin{bmatrix} 1 & -1 & -1 & -1 \\ 5 & -4 & -5 & -5 \\ 4 & 1 & -3 & -4 \\ 1 & 3 & 2 & -1 \end{bmatrix}.$$

3). Determine if the following subsets of R^3 are subspaces or not. Justify your answer.

a). $S = \{(x, y, z) : x + 3y = 5z, \ x, y, z \in R\}$.

b). $S = \{(x, y, z) : x + 3y = 5, \ x, y, z \in R\}$.

c). $S = \{(x,0,z) : x = 3z,\ x,y,z \in R\}$.

d). $S = \{(x,y,z) : xyz = 0,\ x,y,z \in R\}$.

4). For those S's, in (3), that are subspaces find a basis for the subspace.

5). For the matrices in problem 1 find:

a). A basis for the range space.

b). A basis for the null space.

6). Let y = f(x) be a function represented by y = Ax where

$$A = \begin{bmatrix} 1 & 1 & 1 & 0 & 1 \\ 2 & 3 & 2 & -1 & 2 \\ -1 & 1 & -1 & -2 & -1 \end{bmatrix}$$

and x and y are referred to the standard basis. Let the columns of

$$S = \begin{bmatrix} 1 & 0 & 1 \\ -4 & 1 & -3 \\ -4 & -2 & -5 \end{bmatrix}$$

be a new basis for the range of f and let

$$T = \begin{bmatrix} 1 & 1 & -1 & 0 & 1 \\ -4 & -3 & 4 & 1 & -5 \\ 3 & -1 & -2 & -3 & 6 \\ -4 & -6 & 2 & -4 & 0 \\ -2 & -7 & 7 & 0 & -2 \end{bmatrix}$$

be a new basis for the domain of f. Find the new representation of y = f(x) as y = Bx

where now x and y are referred to the new basis.

Index

Note: Maple commands are listed in boldface.